Lectures on Fuzzy and Fuzzy SUSY Physics

Lectures on Fuzzy and Fuzzy SUSY Physics

A. P. Balachandran
Syracuse University, USA

S. Kürkçüoğlu
Dublin Institute for Advanced Studies, Ireland

S. Vaidya
Indian Institute of Science, India

World Scientific

NEW JERSEY · LONDON · SINGAPORE · BEIJING · SHANGHAI · HONG KONG · TAIPEI · CHENNAI

Published by

World Scientific Publishing Co. Pte. Ltd.

5 Toh Tuck Link, Singapore 596224

USA office: 27 Warren Street, Suite 401-402, Hackensack, NJ 07601

UK office: 57 Shelton Street, Covent Garden, London WC2H 9HE

British Library Cataloguing-in-Publication Data
A catalogue record for this book is available from the British Library.

LECTURES ON FUZZY AND FUZZY SUSY PHYSICS

ISBN-13 978-981-270-466-5
ISBN-10 981-270-466-3

Printed in Singapore.

Dedicated to Rafael Sorkin,
our friend, teacher,
a true and creative seeker of knowledge.

Preface

One of us (Balachandran) gave a course of lectures on "Fuzzy Physics" during spring, 2002 for students of Syracuse and Brown Universities. The course which used video conferencing technology was also put on the websites [1]. Subsequently A.P. Balachandran, S. Kürkçüoğlu and S.Vaidya decided to edit the material and publish them as lecture notes. The present book is the outcome of that effort.

The recent interest in fuzzy physics begins from the work of Madore [2, 3] even though the basic physical and mathematical ideas are older and go back to Hoppe [4]. Many of the mathematical developments are based on the works of Kostant and Kirillov [5] and Berezin [6]. They emerge from the fundamental observation that coadjoint orbits of Lie groups are symplectic manifolds which can therefore be quantized under favorable circumstances. When that can be done, we get a quantum representation of the manifold. It is the fuzzy manifold for the underlying "classical manifold". It is fuzzy because no precise localization of points thereon is possible. The fuzzy manifold approaches its classical version when the effective Planck's constant of quantization goes to zero.

Our interest will be in compact simple and semi-simple Lie groups for which coadjoint and adjoint orbits can be identified and are compact as well. In such a case these fuzzy manifold is a finite-dimensional matrix algebra on which the Lie group acts in simple ways. Such fuzzy spaces are therefore very simple and also retain the symmetries of their classical spaces. These are some of the reasons for their attraction.

There are several reasons to study fuzzy manifolds. Our interest has its roots in quantum field theory (qft). Qft's require regularization and the conventional nonperturbative regularization is lattice regularization. It has been extensively studied for over thirty years. It fails to preserve space-time symmetries of quantum fields. It also has problems in dealing with topological subtleties like instantons, and can deal with index theory and axial anomaly only approximately. Instead fuzzy physics does not have these problems. So it merits investigation as an alternative tool to investigate qft's.

A related positive feature of fuzzy physics, is its ability to deal with supersymmetry(SUSY) in a precise manner [7–10]. (See however,[11]). Fuzzy SUSY models are also finite-dimensional matrix models amenable to numerical work, so this is another reason for our attraction to this field.

Interest in fuzzy physics need not just be utilitarian. Physicists have long speculated that space-time in the small has a discrete structure. Fuzzy space-time gives a very concrete and interesting method to model this speculation and test its consequences. There are many generic consequences of discrete space-time, like CPT and causality violations, and distortions of the Planck spectrum. Among these must be characteristic signals for fuzzy physics, but they remain to be identified.

We are a part of a collaboration on fuzzy physics and noncommutative geometry. These lecture notes have been strongly influenced by our interactions with our colleagues in this collaboration. We thank them for discussions and ideas. The work of APB has been supported in part by DOE under contract number DE-FG02-85ER40231. The work of SK is supported by Irish Research Council for Science Engineering and Technology(IRCSET).

These lecture notes are not exhaustive, and reflect the research interests of the authors. It is our hope that the interested reader will be able to learn about the topics we have not covered with the help of our citations.

Contents

Chapter 1

Introduction

We can find few fundamental physical models amenable to exact treatment. Approximation methods like perturbation theory are necessary and are part of our physics culture.

Among the important approximation methods for quantum field theories (qft's) are strong coupling methods based on lattice discretization of the underlying spacetime or perhaps its time-slice. They are among the rare effective approaches for the study of confinement in QCD and for nonperturbative regularization of qft's. They enjoyed much popularity in their early days and have retained their good reputation for addressing certain fundamental problems.

One feature of naive lattice discretizations however can be criticized. They do not retain the symmetries of the exact theory except in some rough sense. A related feature is that topology and differential geometry of the underlying manifolds are treated only indirectly, by limiting the couplings to "nearest neighbors". Thus lattice points are generally manipulated like a trivial topological set, with a point being both open and closed. The upshot is that these models have no rigorous representation of topological defects and lumps like vortices, solitons and monopoles. The complexities in the ingenious lattice representations of the QCD θ-term [12] illustrate such limitations. There do exist radical attempts to overcome these limitations using partially ordered sets [13], but their potentials are yet to be adequately studied.

As mentioned in the preface, a new approach to discretization, under the name of "fuzzy physics" inspired by noncommutative geometry (NCG), is being developed for a while now. The key remark here is that when the underlying spacetime or spatial cut can be treated as a phase space and quantized, with a parameter $\hat{\hbar}$ assuming the role of $\hbar$, the emergent quantum space is fuzzy, and the number of independent states per ("classical") unit volume becomes finite. We have known this result after Planck and Bose introduced such an ultraviolet cut-off and quantum physics later justified it. A "fuzzified" manifold is expected to be ultraviolet finite, and if the parent manifold is compact too, supports only finitely many independent states. The continuum limit is the semi-classical $\hat{\hbar} \to 0$ limit. This unconventional discretization of classical topology is not at all equivalent to the naive one, and we shall see that it does significantly overcome the previous criticisms.

There are other reasons also to pay attention to fuzzy spaces, be they spacetimes or spatial slices. There is much interest among string theorists in matrix models and in describing D-branes using matrices. Fuzzy spaces lead to matrix models too and their ability to reflect topology better than elsewhere should therefore evoke our curiosity. They let us devise new sorts of discrete models and are interesting from that perspective. In addition, as mentioned in the preface, it has now been discovered that when open strings end on D-branes which are symplectic manifolds, then the branes can become fuzzy. In this way one comes across fuzzy tori, $\mathbb{C}P^N$ and many such spaces in string physics.

The central idea behind fuzzy spaces is discretization by quantization. It does not always work. An obvious limitation is that the parent manifold has to be even dimensional. If it is not, it has no chance of being a phase space. But that is not all. Successful use of fuzzy spaces for qft's requires good fuzzy versions of the Laplacian, Dirac equation, chirality operator and so forth, and their incorporation can make the entire enterprise complicated. The torus T^2 is compact, admits a symplectic structure and on quantization becomes a fuzzy, or a non-commutative torus. It supports a finite number of states if the symplectic form satisfies the Dirac quantization condition. But it is impossible to introduce suitable derivations without escalating the formalism to infinite dimensions.

But we do find a family of classical manifolds elegantly escaping these limitations. They are the co-adjoint orbits of Lie groups. For semi-simple Lie groups, they are the same as adjoint orbits. It is a theorem that these orbits are symplectic. They can often be quantized when the symplectic forms satisfy the Dirac quantization condition. The resultant fuzzy spaces are described by linear operators on irreducible representations (IRR's) of the group. For compact orbits, the latter are finite-dimensional. In addition, the elements of the Lie algebra define natural derivations, and that helps to find the Laplacian and the Dirac operator. We can even define chirality with no fermion doubling and represent monopoles and instantons. (See chapters 5, 6 and 8). These orbits therefore are altogether well-adapted for qft's.

Let us give examples of these orbits:

- $S^2 \simeq \mathbb{C}P^1$: This is the orbit of $SU(2)$ through the Pauli matrix σ_3 or any of its multiples $\lambda\sigma_3$ ($\lambda \neq 0$). It is the set $\{\lambda\,g\,\sigma_3\,g^{-1} : g \in SU(2)\}$. The symplectic form is $j\,d\,(\cos)\theta \wedge d\phi$ with θ, ϕ being the usual S^2 coordinates. Quantization gives the spin j $SU(2)$ representations.

- $\mathbb{C}P^2$: $\mathbb{C}P^2$ is of particular interest being of dimension 4. It is the orbit of $SU(3)$ through the hypercharge $Y = 1/3\,\mathrm{diag}(1,1,-2)$ (or its non-zero multiples):

$$\mathbb{C}P^2 : \{g\,Y\,g^{-1} : g \in SU(3)\}. \tag{1.1}$$

  The associated representations are symmetric products of 3's or $\bar{3}$'s.

  In a similar way $\mathbb{C}P^N$ are adjoint orbits of $SU(N+1)$ for any $N \geq 3$. They too can be quantized and give rise to fuzzy spaces.

- $SU(3)/[U(1) \times U(1)]$: This 6-dimensional manifold is the orbit of $SU(3)$ through $\lambda_3 = \mathrm{diag}(1,-1,0)$ and its non-zero multiples. These orbits give all the IRR's containing a zero hypercharge state.

In this book, we focus on the fuzzy spaces emerging from quantizing S^2. They are called the fuzzy spheres S^2_F and depend on the integer or half integer j labelling the irreducible representations of $SU(2)$. Physics on S^2_F is treated in detail. Scalar and gauge fields, the Dirac operator, instantons, index theory, and the so-called UV-IR mixing [14–20] are all covered. Supersymmetry can be elegantly discretized in the approach of fuzzy physics by replacing the Lie algebra $su(2)$ of $SU(2)$ by the superalgebras $osp(2,1)$ and $osp(2,2)$. Fuzzy supersymmetry is also discussed in these lectures including its instanton and index theories. We also briefly discuss the fuzzy spaces associated with $\mathbb{C}P^N$ ($N \geq 2$). These spaces, especially $\mathbb{C}P^2$, are of physical interest. We refer to the literature [29–33] for their more exhaustive treatment.

Fuzzy physics draws from many techniques and notions developed in the context of noncommutative geometry. There are excellent books and reviews on this vast subject some of which we include in the bibliography [3, 34–38].

Chapter 2

Fuzzy Spaces

In the present chapter, we approach the problem of quantization of classical manifolds like S^2 and $\mathbb{C}P^N$ using harmonic oscillators. The method is simple and transparent, and enjoys generality too. The point of departure in this approach is the quantization of complex planes. We focus on quantizing $\mathbb{C}^2$ and its associated S^2 first. We will consider other manifolds later in the chapter.

2.1 Fuzzy $\mathbb{C}^2$

The two-dimensional complex plane $\mathbb{C}^2$ has coordinates $z = (z_1, z_2)$ where $z_i \in \mathbb{C}$. We want to quantize $\mathbb{C}^2$ turning it into fuzzy $\mathbb{C}^2 \equiv \mathbb{C}_F^2$.

This is easily accomplished. After quantization, z_i become harmonic oscillator annihilation operators a_i and z_i^* become their adjoint. Their commutation relations are

$$[a_i \, a_j] = [a_i^\dagger \, a_j^\dagger] = 0 \,, \qquad [a_i \, a_j^\dagger] = \tilde{\hbar} \delta_{ij} \,, \tag{2.1}$$

where the $\tilde{\hbar}$ need not be the "Planck's constant/2π". The classical manifold emerges as $\tilde{\hbar} \to 0$. We set the usual Planck's constant $\hbar$ to 1 hereafter unless otherwise stated.

In the same way, we can quantize $\mathbb{C}^{N+1}$ for any N using an appropriate number of oscillators and that gives us fuzzy $\mathbb{C}P^N$ as we shall later see.

2.2 Fuzzy S^3 and Fuzzy S^2

There is a well-known descent chain from $\mathbb{C}^2$ to the 3-sphere S^3 and thence to S^2. Our tactic to obtain fuzzy $S^2 \equiv S_F^2$ is to quantize this chain, obtaining along the way fuzzy $S^3 \equiv S_F^3$.

Let us recall this chain of manifolds. Consider $\mathbb{C}^2$ with the origin removed, $\mathbb{C}^2 \setminus \{0\}$. As $z \neq 0$, $\frac{z}{|z|}$ with $|z| = \left(\sum_i |z_i|^2 \right)^{\frac{1}{2}}$ makes sense here. Since $\frac{z}{|z|}$ is normalized to 1, $\left| \frac{z}{|z|} \right| = 1$, it gives the 3-sphere S^3. Thus we have the fibration

$$\mathbb{R} \to \mathbb{C}^2 \setminus \{0\} \to S^3 = \left\langle \xi = \frac{z}{|z|} \right\rangle , \qquad z \to \frac{z}{|z|} . \tag{2.2}$$

Now S^3 is a $U(1)$-bundle ("Hopf fibration") [45] over S^2. If $\xi \in S^3$, then $\vec{x}(\xi) = \xi^\dagger \vec{\tau} \xi$ (where τ_i, $i = 1, 2, 3$ are the Pauli matrices) is invariant under the $U(1)$ action $\xi \to \xi e^{i\theta}$ and is a real normalized three-vector:

$$\vec{x}(\xi)^* = \vec{x}(\xi) , \qquad \vec{x}(\xi) \cdot \vec{x}(\xi) = 1 . \tag{2.3}$$

So $\vec{x}(\xi) \in S^2$ and we have the Hopf fibration

$$U(1) \to S^3 \to S^2 , \qquad \xi \to \vec{x}(\xi) . \tag{2.4}$$

Note that $\vec{x}(\xi) = \frac{1}{|z|} z^* \vec{\tau} z \frac{1}{|z|}$.

The fuzzy S^3 is obtained by replacing $\frac{z_i}{|z|}$ by $a_i \frac{1}{\sqrt{\hat{N}}}$ where $\hat{N} = \sum_j a_j^\dagger a_j$ is the number operator:

$$\frac{z_i}{|z|} \to a_i \frac{1}{\sqrt{\hat{N}}} , \qquad \frac{z_i^*}{|z|} \to \frac{1}{\sqrt{\hat{N}}} a_i^\dagger , \qquad \hat{N} = \sum_j a_j^\dagger a_j , \, \hat{N} \neq 0 . \tag{2.5}$$

The quantum condition $\hat{N} \neq 0$ means that the vacuum is omitted from the Hilbert space, so that it is the orthogonal complement of the vacuum in Fock space. This omission is like the deletion of 0 from $\mathbb{C}^2$.

There is a problem with this omission as $a_i \frac{1}{\sqrt{\hat{N}}}$ and its polynomials will create it from any $\hat{N} = n$ state. For this reason, and because $a_i \frac{1}{\sqrt{\hat{N}}}$ and its adjoint need the infinite-dimensional Fock space to act on and do not give finite-dimensional models for S_F^3, we will not dwell on this space.

2.3   The Fuzzy Sphere S_F^2

The problems of S_F^3 melt away for S_F^2. Quantization of S^2 gives S_F^2 with $x_i(\xi)$ becoming the operator x_i:

$$x_i(\xi) \to x_i = \frac{1}{\sqrt{\hat{N}}} a^\dagger \vec{\tau} a \frac{1}{\sqrt{\hat{N}}} = \frac{1}{\hat{N}} a^\dagger \tau_i a , \qquad \hat{N} \neq 0 . \tag{2.6}$$

Since

$$[x_i , \hat{N}] = 0 , \tag{2.7}$$

we can restrict x_i to the subspace $\mathcal{H}_n$ of the Fock space where $\hat{N} = n \, (\neq 0)$. This space is $(n+1)$-dimensional and is spanned by the orthonormal vectors

$$\frac{(a_1^\dagger)^{n_1}}{\sqrt{n_1!}} \frac{(a_2^\dagger)^{n_2}}{\sqrt{n_2!}} |0\rangle \equiv |n_1 \, n_2\rangle \,, \quad n_1 + n_2 = n \,. \tag{2.8}$$

The x_i act irreducibly on this space and generate the full matrix algebra $Mat(n+1)$.

The $SU(2)$ angular momentum operators L_i are given by the Schwinger construction:

$$L_i = a^\dagger \frac{\tau_i}{2} a \,, \quad [L_i \,, L_j] = i\epsilon_{ijk}L_k \,. \tag{2.9}$$

$a_i^\dagger$ transform as spin $\frac{1}{2}$ spinors and (2.8) spans the n-fold symmetric product of these spinors. It has angular momentum $\frac{n}{2}$:

$$L_i L_i \big|_{\mathcal{H}_n} = \frac{n}{2}\left(\frac{n}{2} + 1\right)\mathbf{1}\big|_{\mathcal{H}_n} \,. \tag{2.10}$$

Since

$$x_i \big|_{\mathcal{H}_n} = \frac{2}{n} L_i \big|_{\mathcal{H}_n} \,, \tag{2.11}$$

we find

$$[x_i \,, x_j] \big|_{\mathcal{H}_n} = \frac{2}{n} i\epsilon_{ijk} x_k \big|_{\mathcal{H}_n} \,,$$

$$\left(\sum_i x_i^2\right)\big|_{\mathcal{H}_n} = \left(1 + \frac{2}{n}\right)\mathbf{1}\big|_{\mathcal{H}_n} \,. \tag{2.12}$$

S_F^2 has radius $\left(1 + \frac{2}{n}\right)^{\frac{1}{2}}$ which becomes 1 as $n \to \infty$.

We generally write the equations in (2.12) as $[x_i \,, x_j] = \frac{2}{n} i\epsilon_{ijk} x_k$, $\left(\sum_i x_i^2\right) = \left(1 + \frac{2}{n}\right)$, omitting the indication of $\mathcal{H}_n$. S_F^2 should have an additional label n, but that too is usually omitted. The x_i's are seen to commute in the naive continuum limit $n \to \infty$ giving back the commutative algebra of functions on S^2.

The fuzzy sphere S_F^2 is a "quantum" object. It has wave functions which are generated by x_i restricted to $\mathcal{H}_n$. Its Hilbert space is $Mat(n+1)$ with the scalar product

$$(m_1, m_2) = \frac{1}{n+1} Tr \, m_1^\dagger m_2 \,, \quad m_i \in Mat(n+1) \,. \tag{2.13}$$

We denote $Mat(n+1)$ with this scalar product also as $Mat(n+1)$.

2.4   Observables of S_F^2

The observables of S_F^2 are associated with linear operators on $Mat(n+1)$.
We can associate two linear operators α^L and α^R to each $\alpha \in Mat(n+1)$.
They have left- and right-actions on $Mat(n+1)$;

$$\alpha^L m = \alpha m\,, \qquad \alpha^R m = m\alpha\,, \qquad \forall \quad m \in Mat(n+1) \tag{2.14}$$

and fulfill

$$(\alpha\beta)^L = \alpha^L \beta^L\,, \qquad (\alpha\beta)^R = \beta^R \alpha^R\,. \tag{2.15}$$

Such left- and right- operators commute:

$$[\alpha^L, \beta^R] = 0\,, \qquad \forall \quad \alpha, \beta \in Mat(n+1)\,. \tag{2.16}$$

We denote the two commuting matrix algebras of left- and right- operators
by $Mat_{L,R}(n+1)$. $Mat(n+1)$ is generated by $a_i^\dagger a_j$ with the understanding
that their domain is $\mathcal{H}_n$. Accordingly, $Mat_{L,R}(n+1)$ are generated by
$(a_i^\dagger a_j)^{L,R}$.

We can also define operators $a_i^{L,R}$, $(a_j^\dagger)^{L,R}$:

$$\begin{aligned}
a_i^L m &= a_i m\,, & a_i^R m &= m a_i \\
a_j^{\dagger L} m &= a_j^\dagger m\,, & a_j^{\dagger R} m &= m a_j^\dagger\,.
\end{aligned} \tag{2.17}$$

They are operators changing n:

$$\begin{aligned}
a_i^{L,R} &: & \mathcal{H}_n &\to \mathcal{H}_{n-1} \\
a_j^{\dagger L,R} &: & \mathcal{H}_n &\to \mathcal{H}_{n+1}\,.
\end{aligned} \tag{2.18}$$

Such operators are important for discussions of bundles. (See chapter 5.)
With the help of these operators, we can write

$$(a_i^\dagger a_j)^L = a_i^{\dagger L} a_j^L\,, \qquad (a_i^\dagger a_j)^R = a_j^R a_i^{\dagger R}\,. \tag{2.19}$$

Of particular interest are the three angular momentum operators

$$L_i^L, L_i^R\,, \qquad \mathcal{L}_i = L_i^L - L_i^R\,. \tag{2.20}$$

Of these, $\mathcal{L}_i$ annihilates $\mathbf{1}$ as does the continuum orbital angular momentum. It is the fuzzy sphere angular momentum approaching the orbital
angular momentum of S^2 as $n \to \infty$:

$$\mathcal{L}_i \to -i\big(\vec{x}(\xi) \wedge \vec{\nabla}\big) \equiv -i\epsilon_{ijk} x(\xi)_j \frac{\partial}{\partial x(\xi)_k} \quad \text{as} \quad n \to \infty\,. \tag{2.21}$$

2.5 Diagonalizing $\mathcal{L}^2$

We have $\sum_i (L_i^L)^2 = \sum_i (L_i^R)^2 = \frac{n}{2}\left(\frac{n}{2}+1\right)$ so that orbital angular momentum is the sum of two angular momenta with values $\frac{n}{2}$. Hence the spectrum of $\mathcal{L}^2$ is

$$\langle \ell(\ell+1) : \ell \in \{0, 1, 2, ..., n\}\rangle. \tag{2.22}$$

A function f in $C^\infty(S^2)$ has the expansion

$$f = \sum_{\ell, m} a_{\ell m} Y_{\ell m} \tag{2.23}$$

in terms of the spherical harmonics $Y_{\ell m}$. The spectrum of orbital angular momentum is thus $\langle \ell(\ell+1) : \ell \in \{0, 1, 2, \ldots, n, , \ldots\}\rangle$.

The spectrum of $\mathcal{L}^2$ is thus precisely that of the continuum orbital angular momentum cut off at n. There is no distortion of eigenvalues upto n.

The eigenstates T_m^ℓ, $m \in \{-\ell, -\ell+1, ..., \ell\}$ of $\mathcal{L}^2$ are known as polarization tensors [46]. They are eigenstates of $\mathcal{L}_3$ and also orthonormal:

$$\mathcal{L}^2 T_m^\ell = \ell(\ell+1) T_m^\ell ,$$
$$\mathcal{L}_3 T_m^\ell = m T_m^\ell ,$$
$$\left(T_{m'}^{\ell'}, T_m^\ell\right) = \delta_{\ell\ell'} \delta_{m'm} . \tag{2.24}$$

2.6 Scalar Fields on S_F^2

We will be brief here as they are treated in detail in chapter 4. A complex scalar field Φ on S^2 is a power series in the coordinate functions $m_i := x_i$,

$$\Phi = \sum a_{i_1 \ldots i_n} m_{i_1} \cdots m_{i_n} . \tag{2.25}$$

(Note again that $\vec{m} \cdot \vec{m} = 1$) The Laplacian on S^2 is $\Delta := -(-i\vec{x} \wedge \vec{\nabla})^2$ and a simple Euclidean action is

$$S = -\int \frac{d\Omega}{4\pi} \Phi^* \Delta \Phi \qquad d\Omega = d\cos(\theta) d\phi . \tag{2.26}$$

We can simplify (2.26) by the expansion

$$\Phi(\vec{x}) = \sum \Phi_{\ell m} Y_{\ell m}(\vec{m}) . \tag{2.27}$$

Then since $\Delta Y_{\ell m}(\vec{m}) = -\ell(\ell+1) Y_{\ell m}(\vec{m})$, and $\int \frac{d\Omega}{4\pi} Y_{\ell'm'}(\vec{m})^* Y_{\ell m}(\vec{m}) = \delta_{\ell\ell'}\delta_{mm'}$,

$$S = \sum_{\ell, m} \ell(\ell+1) \Phi_{\ell m}^* \Phi_{\ell m} . \tag{2.28}$$

From (2.25), we infer that the fuzzy scalar field Ψ is a power series in the matrices x_i and is hence itself a matrix. The Euclidean action replacing (2.26) is

$$S = (\mathcal{L}_i\Psi \,, \mathcal{L}_i\Psi) = -(\Psi \,, \Delta\Psi) \,. \tag{2.29}$$

On expanding ψ according to

$$\Psi = \sum_{\ell \leq n+1} \psi_{\ell m} T_m^\ell \,, \tag{2.30}$$

this reduces to

$$S = \sum_{\ell \leq n} \ell(\ell+1)|\psi_{\ell m}|^2 \,. \tag{2.31}$$

2.7 The Holstein-Primakoff Construction

There is an interesting construction of L_i for fixed n using just one oscillator due to Holstein and Primakoff [47]. We outline this construction here.

In brief, since $\hat{N}$ commutes with L_i, we can eliminate a_2 from L_i and restrict L_i to $\mathcal{H}_n$ without spoiling their commutation relations. The result is the Holstein-Primakoff construction.

We now give the details. (2.5) gives the following polar decomposition of a_2:

$$a_2 = U\sqrt{N - a_1^\dagger a_1} \,, \qquad U^\dagger U = UU^\dagger = \mathbf{1} \,, \tag{2.32}$$

where we choose the positive square root:

$$\sqrt{N - a_1^\dagger a_1} \geq 0 \,. \tag{2.33}$$

We can understand U better by examining the action of a_2 on the orthonormal states (2.8) spanning $\mathcal{H}_n$. We find

$$\begin{aligned}
a_2|n_1 \,, n_2\rangle &= \sqrt{n_2}|n_1 \,, n_2 - 1\rangle \\
&= U\sqrt{N - a_1^\dagger a_1}|n_1 \,, n_2\rangle \\
&= \sqrt{n_2}U|n_1 \,, n_2\rangle
\end{aligned} \tag{2.34}$$

or

$$U|n_1 \,, n_2\rangle = |n_1 \,, n_2 - 1\rangle \tag{2.35}$$

Thus if

$$A^\dagger = a_1^\dagger U \,, \qquad A = U^\dagger a_1 \,, \tag{2.36}$$

then

$$A^\dagger |n_1, n_2\rangle = \sqrt{n_1 + 1}|n_1 + 1, n_2 - 1\rangle,$$
$$A|n_1, n_2\rangle = \sqrt{n_1}|n_1 - 1, n_2 - 1\rangle, \qquad (2.37)$$

and

$$[A, A^\dagger] = \mathbf{1}, \quad [A^\dagger, A^\dagger] = [A, A] = 0,$$
$$[A, N] = [A^\dagger, N] = 0. \qquad (2.38)$$

a_2 vanishes on $|n, 0\rangle$ and U and $A^\dagger$ are undefined on that vector. That is, A and $A^\dagger$ can not be defined on $\mathcal{H}_n$. In any case, the oscillator algebra of (2.38) has no finite-dimensional representation. But this is not the case for L_i. We have

$$L_+ = L_1 + iL_2 = a_1^\dagger a_2 = A^\dagger \sqrt{N - A^\dagger A}$$
$$L_- = L_1 - iL_2 = a_2^\dagger a_1 = \sqrt{N - A^\dagger A}\, A$$
$$L_3 = a_1^\dagger a_1 - a_2^\dagger a_2 = A^\dagger A - N \qquad (2.39)$$

On $\mathcal{H}_n$, (2.39) gives the Holstein-Primakoff realization of the $SU(2)$ Lie algebra for angular momentum $\frac{n}{2}$.

2.8 $\mathbb{C}P^N$ and Fuzzy $\mathbb{C}P^N$

S^2 is $\mathbb{C}P^1$ as a complex manifold. The additional structure for $\mathbb{C}P^1$ as compared to S^2 is only the complex structure. So we can without great harm denote S^2 and S_F^2 also as $\mathbb{C}P^1$ and $\mathbb{C}P_F^1$. In section 5.4.2, we will in fact consider the complex structure and its quantization.

Generalizations of $\mathbb{C}P^1$ and $\mathbb{C}P_F^1$ are $\mathbb{C}P^N$ and $\mathbb{C}P_F^N$. They are associated with the groups $SU(N+1)$.

Classically $\mathbb{C}P^N$ is the complex projective space of complex dimension N. It can described as follows. Consider the $(2N + 1)$-dimensional sphere

$$S^{2N+1} = \left\langle \xi = \left(\xi_1, \xi_2 \cdots, \xi_{N+1}\right) : \xi_i \in \mathbb{C}, |\xi|^2 := \sum |\xi_i|^2 = 1 \right\rangle. \quad (2.40)$$

It admits the $U(1)$ action

$$\xi \to e^{i\theta}\xi. \qquad (2.41)$$

$\mathbb{C}P^N$ is the quotient of S^{2N+1} by this action giving rise to the fibration

$$U(1) \to S^{2N+1} \to \mathbb{C}P^N. \qquad (2.42)$$

If λ_i are the Gell-Mann matrices of $SU(N+1)$, a point of $\mathbb{C}P^N$ is

$$\vec{X}(\xi) = \xi^\dagger \vec{\lambda} \xi, \quad \xi \in S^{2N+1}. \qquad (2.43)$$

For $N = 1$, these become the previously constructed structures.

There is another description of S^{2N+1} and $\mathbb{C}P^N$. $SU(N+1)$ acts transitively on S^{2N+1} and the stability group at $(1, \vec{0})$ is

$$SU(N) = \left\langle u \in SU(N+1) : u = \begin{pmatrix} 1 & 0 \\ 0 & \hat{u} \end{pmatrix} \right\rangle . \qquad (2.44)$$

Hence

$$S^{2N+1} = SU(N+1)/SU(N) . \qquad (2.45)$$

Consider the equivalence class

$$\langle (1, \vec{0}) \rangle = \langle (e^{i\theta}, \vec{0}) | e^{i\theta} \in U(1) \rangle \qquad (2.46)$$

of all elements connected to $(1, \vec{0})$ by the $U(1)$ action (2.41). Its orbit under $SU(N+1)$ is $\mathbb{C}P^N$. The stability group of (2.46) is

$$S[U(1) \times U(N)] = \left[v \in SU(N+1) : v = \begin{pmatrix} e^{i\theta} & 0 \\ 0 & \hat{v} \end{pmatrix} \right] . \qquad (2.47)$$

Thus

$$\mathbb{C}P^N = SU(N+1)/S[U(1) \times U(N)] . \qquad (2.48)$$

$S[U(1) \times U(N)]$ is commonly denoted as $U(N)$. The two groups are isomorphic.

To obtain $\mathbb{C}P_F^N$, we think of S^{2N+1} as a submanifold of $\mathbb{C}^{N+1} \setminus \{0\}$:

$$S^{2N+1} = \left\langle \xi = \frac{z}{|z|} : z = (z_1, z_2, \cdots, z_{N+1}) \in \mathbb{C}^{N+1} \setminus \{0\} \right\rangle . \qquad (2.49)$$

Just as before, we can quantize $\mathbb{C}^{N+1}$ by replacing z_i by annihilation operators a_i and z_i^* by $a_i^\dagger$:

$$[a_i, a_j] = [a_i^\dagger, a_j^\dagger] = 0 , \quad [a_i, a_j^\dagger] = \delta_{ij} . \qquad (2.50)$$

With

$$\widehat{N} = a_i^\dagger a_i \qquad (2.51)$$

as the number operator, the quantized ξ is given by the correspondence

$$\xi_i = \frac{z_i}{|z|} \longrightarrow a_i \frac{1}{\sqrt{\widehat{N}}} , \quad N \neq 0 . \qquad (2.52)$$

Then as in (2.6), we get the $\mathbb{C}P_F^N$ coordinates

$$X_i(z) \longrightarrow X_i = \frac{1}{\widehat{N}} a^\dagger \lambda_i a , \quad N \neq 0 . \qquad (2.53)$$

The rest of the discussion follows that of $\mathbb{C}P^1$ with $SU(N+1)$ replacing $SU(2)$. Because of (the analogue of) (2.7), X_i can be restricted to $\mathcal{H}_n$, the subspace of the Fock space with $\hat{N} = n$. It is spanned by the orthonormal vectors

$$\prod_{i=1}^{N+1} \frac{(a_i^\dagger)^{n_i}}{\sqrt{n_i!}} |0\rangle := |n_1 \, n_2, \cdots, N+1\rangle, \qquad \sum n_i = n, \qquad (2.54)$$

and is of dimension

$$M = {}^{N+1+n}C_n = \frac{(N+n)!}{n!N!}. \qquad (2.55)$$

The $SU(N+1)$ angular momentum operators are given by a generalized Schwinger construction :

$$L_i = a^\dagger \frac{\lambda_i}{2} a, \qquad [L_i, L_j] = i f_{ijk} L_k. \qquad (2.56)$$

$a_i^\dagger$ transform by the unitary irreducible representation (UIR) of dimension $(N+1)$ of $SU(N+1)$ and (2.54) span the space of n fold symmetric product of these UIR's. It carries a UIR of dimension (2.55) and has the quadratic Casimir operator

$$\sum L_i^2 = \frac{N}{2} \left(\frac{n^2}{N+1} + n \right) \mathbf{1}. \qquad (2.57)$$

Its remaining Casimir operators are fixed by (2.57). As before

$$X_i \big|_{\mathcal{H}_n} = \frac{2}{n} L_i \big|_{\mathcal{H}_n},$$

$$[X_i, X_j] \big|_{\mathcal{H}_n} = \frac{2}{n} i f_{ijk} X_k \big|_{\mathcal{H}_n},$$

$$\left(\sum X_i^2 \right) \big|_{\mathcal{H}_n} = \left(\frac{2N}{N+1} + \frac{2N}{n} \right) \mathbf{1} \big|_{\mathcal{H}_n}. \qquad (2.58)$$

The "size" of $\mathbb{C}P_F^N$ is measured by the "radius" $\sqrt{\left(\frac{2N}{N+1} + \frac{2N}{n} \right)}$. In the $n \to \infty$ limit, the X_i's also commute and generate $C^\infty(\mathbb{C}P^N)$.

The "wave functions" of $\mathbb{C}P_F^N$ are polynomials in X_i, that is they are elements of $Mat(M)$, with a scalar product like (2.13). As before, for each $\alpha \in Mat(M)$, we have two observables $\alpha_{L,R}$ and they constitute the matrix algebras $Mat_{L,R}(M)$.

The discussions leading up to (2.18) and (2.20) can be adapted also to $\mathbb{C}P_F^N$. As for (2.21), it generalizes to

$$\mathcal{L}_i \longrightarrow -i f_{ijk} X_j(\xi) \frac{\partial}{\partial X_k(\xi)}. \qquad (2.59)$$

Diagonalization of $\mathcal{L}_i$ involves the reduction of the product of the UIR's of $SU(N+1)$ given by L_i^L and its complex conjugate given by L_i^R to their irreducible components. The corresponding polarization operators can also in principle be constructed.

The scalar field action (2.28) generalizes easily to $\mathbb{C}P_F^N$.

2.9 The $\mathbb{C}P^N$ Holstein-Primakoff Construction

The generalization of this construction to $\mathbb{C}P^N$ and $SU(N+1)$ is due to Sen [48].

Consider for specificity $N = 2$ and $SU(3)$ first. $SU(3)$ has 3 oscillators a_1, a_2, a_3. There are also the $SU(2)$ algebras with generators

$$\sum_{i=j=1}^{2} a_i^\dagger \left(\frac{\vec{\sigma}}{2}\right)_{ij} a_j , \qquad \sum_{i,j=2}^{3} a_i^\dagger \left(\frac{\vec{\sigma}}{2}\right)_{ij} a_j , \qquad (2.60)$$

acting on the indices $1, 2$ and $2, 3$ respectively, of a's and $a^\dagger$'s. Taking their commutators, we can generate the full $SU(3)$ Lie algebra.

We will eliminate a_2, $a_2^\dagger$ from both these sets using the previous Holstein-Primakoff construction. In that way, we will obtain the $SU(3)$ Holstein-Primakoff construction.

As previously we write the polar decompositions

$$a_2 = U_2\sqrt{N_2} , \quad a_2^\dagger = \sqrt{N_2}U_2^\dagger , \quad N_2 = a_2^\dagger a_2 , \quad U_2^\dagger U_2 = \mathbf{1} . \qquad (2.61)$$

The oscillators act on the Fock space $\oplus_N \mathcal{H}_N$ spanned by (2.54) for $N = 2$. The actions of U_2 and

$$A_{12}^\dagger = a_1^\dagger U_2 , \quad A_{12} = U_2^\dagger a_1 \qquad (2.62)$$

follow (2.35) and (2.36). They do not affect n_3. Using (2.39), we can write the $SU(2)$ generators acting on (12) indices as

$$I_+ = a_1^\dagger a_2 = A_{12}^\dagger \sqrt{N_2} ,$$
$$I_- = a_2^\dagger a_1 = \sqrt{N_2} A_{12} ,$$
$$I_3 = \frac{1}{2}(a_1^\dagger a_1 - a_2^\dagger a_2) = \frac{1}{2}(A_{12}^\dagger A_{12} - N_2) . \qquad (2.63)$$

We follow the I, U, V spin notation of $SU(3)$ in particle physics [49]. They are connected by Weyl reflections.

In a similar manner, the $SU(3)$ generators acting on (23) indices are constructed from

$$A_{32}^\dagger = a_3^\dagger U_2 , \quad A_{32} = U_2^\dagger a_3 , \qquad (2.64)$$

and read

$$U_+ = a_3^\dagger a_2 = A_{32}^\dagger \sqrt{N_2}\,,$$
$$U_- = a_2^\dagger a_3 = \sqrt{N_2} A_{32}\,,$$
$$U_3 = \frac{1}{2}\left(a_2^\dagger a_2 - a_3^\dagger a_3\right) = \frac{1}{2}\left(N_2 - A_{32}^\dagger A_{32}\right). \tag{2.65}$$

In a UIR of $SU(3)$, the total number operator $N = N_1 + N_2 + N_3$ is fixed. Acting on $\mathcal{H}_n$, it becomes n. Keeping this in mind, we now substitute

$$N_2 = N - N_1 - N_3 = N - A_{12}^\dagger A_{12} - A_{32}^\dagger A_{32} \tag{2.66}$$

in (2.63) and (2.65) to eliminate the second oscillator. That gives

$$I_+ = A_{12}^\dagger \sqrt{N - N_1 - N_3}\,,$$
$$I_- = \sqrt{N - N_1 - N_3} A_{12}\,,$$
$$I_3 = N_1 + \frac{N_3}{2} - \frac{N}{2} \tag{2.67}$$
$$U_+ = A_{32}^\dagger \sqrt{N - N_1 - N_3}\,,$$
$$U_- = \sqrt{N - N_1 - N_2} A_{32}\,,$$
$$U_3 = -\left(N_3 + \frac{N_1}{2} - \frac{N}{2}\right) \tag{2.68}$$

These operators and their commutators generate the full $SU(3)$ Lie algebra when restricted to $\mathcal{H}_n$. That is the $SU(3)$ Holstein-Primakoff construction.

If the restriction to $\mathcal{H}_n$ is not made, N is a new operator and we get instead the $U(3)$ Lie algebra with N generating its central $U(1)$.

The Holstein-Primakoff construction for $\mathbb{C}P^N$ is much the same. One introduces $N+1$ oscillators $a_i\,, a_i^\dagger (i \in [1, \cdots N])$ with which the $SU(N+1)$ Lie algebra can be realized using the Schwinger construction. The $SU(N+1)$ UIR's we get therefrom are symmetric products of the fundamental representation of dimension $(N+1)$. The number operator $N = a^\dagger \cdot a$ has a fixed value in one such UIR. Next $a_2\,, a_2^\dagger$ are eliminated from $SU(N+1)$ generators in favor of the N and the remaining operators to obtain the generalized Holstein-Primakoff construction.

$SU(N+1)$ is of rank N, and we can realize its Lie algebra with N oscillators. There is a similar result in quantum field theory where with the help of the vertex operator construction, a (simply laced) rank N Lie algebra can be realized with N scalar fields on $S^1 \times \mathbb{R}$ valued on S^1 [50]. This resemblance perhaps is not an accident.

Chapter 3

Star Products

3.1 Introduction

The algebra of smooth functions on a manifold $\mathcal{M}$ under point-wise multiplication is commutative. In deformation quantization [51], this point-wise product is deformed to a non-commutative (but still associative) product called the $*$-product. It has a central role in many discussions of non-commutative geometry. It has been fruitfully used in quantum optics for a long time.

The existence of such deformations was understood many years ago by Weyl, Wigner, Groenewold and Moyal [52, 40, 43]. They noted that if there is a linear injection (one-to-one map) ψ of an algebra $\mathcal{A}$ into smooth functions $\mathbb{C}^\infty(\mathcal{M})$ on a manifold $\mathcal{M}$, then the product in $\mathcal{A}$ can be transported to the image $\psi(\mathcal{A})$ of $\mathcal{A}$ in $\mathbb{C}^\infty(\mathcal{M})$ using the map. That is then a $*$-product.

Let us explain this construction with greater completeness and generality [31]. For concreteness we can consider $\mathcal{A}$ to be an algebra of bounded operators on a Hilbert space closed under the Hermitian conjugation of $*$. It is then an example of a $*$-algebra.

More generally, $\mathcal{A}$ can be a generic "$*$-algebra', that is an algebra closed under an anti-linear involution:

$$a\,,b \in \mathcal{A}\,,\ \lambda \in \mathbb{C} \Rightarrow a^*,b^* \in \mathcal{A}\,,\ (ab)^* = b^*a^*\,,\ (\lambda a)^* = \lambda^* a^*\,. \qquad (3.1)$$

A two-sided ideal $\mathcal{A}_0$ of $\mathcal{A}$ is a subalgebra of $\mathcal{A}$ with the property

$$a_0 \in \mathcal{A}_0 \Rightarrow \alpha a_0 \text{ and } a_0\alpha \in \mathcal{A}_0\,,\ \forall \alpha \in \mathcal{A}\,. \qquad (3.2)$$

That is $\mathcal{A}\mathcal{A}_0\,,\mathcal{A}_0\mathcal{A} \subseteq \mathcal{A}_0$. A two-sided $*$-ideal $\mathcal{A}_0$ by definition is itself closed under $*$ as well.

An element of the quotient $\mathcal{A}/\mathcal{A}_0$ is the equivalence class

$$\{\alpha + \mathcal{A}_0 \subset \mathcal{A}\} = \{[\alpha + a_0]\big| a_0 \in \mathcal{A}_0\}. \tag{3.3}$$

If $\mathcal{A}_0$ is a two-sided ideal, $\mathcal{A}/\mathcal{A}_0$ is itself an algebra with the sum and the product

$$(\alpha + \mathcal{A}_0) + (\beta + \mathcal{A}_0) = \alpha + \beta + \mathcal{A}_0,$$
$$(\alpha + \mathcal{A}_0)(\beta + \mathcal{A}_0) = \alpha\beta + \mathcal{A}_0. \tag{3.4}$$

If $\mathcal{A}_0$ is a two-sided $*$-ideal, then $\mathcal{A}/\mathcal{A}_0$ is a $*$-algebra with the $*$-operation

$$(\alpha + \mathcal{A}_0)^* = \alpha^* + \mathcal{A}_0. \tag{3.5}$$

We note that the product and $*$ are independent of the choice of the representatives α, β from the equivalence classes $\alpha + \mathcal{A}_0$ and $\beta + \mathcal{A}_0$ because $\mathcal{A}_0$ is a two-sided ideal. So they make sense for $\mathcal{A}/\mathcal{A}_0$.

Let $C^\infty(\mathcal{M})$ denote the complex-valued smooth functions on a manifold $\mathcal{M}$. Complex conjugation $-$(bar) is defined on these functions. It sends a function f to its complex conjugate $\bar{f}$.

We consider the linear maps

$$\psi : \mathcal{A} \longrightarrow C^\infty(\mathcal{M}), \tag{3.6}$$

$$\psi\left(\sum \lambda_i a_i\right) = \sum \lambda_i \psi(a_i), \quad a_i \in \mathcal{A}, \quad \lambda_i \in \mathbb{C}. \tag{3.7}$$

The kernel of such a map is the set of all $\alpha \in \mathcal{A}$ for which $\psi(\alpha)$ is the zero function 0. (Its value is zero at all points of $\mathcal{M}$):

$$Ker\,\psi = \langle \alpha_0 \in \mathcal{A}\big| \psi(\alpha_0) = 0 \rangle. \tag{3.8}$$

ψ descends to a linear map, called Ψ, from $\mathcal{A}/Ker\,\psi = \{\alpha + Ker\,\psi : \alpha \in \mathcal{A}\}$ to $C^\infty(\mathcal{M})$:

$$\Psi(\alpha + Ker\,\psi) = \psi(\alpha). \tag{3.9}$$

$\psi(\alpha)$ does not depend on the choice of the representative α from $\alpha + Ker\,\psi$ because of (3.8). Clearly Ψ is an injective map from $\mathcal{A}/Ker\,\psi$ to $C^\infty(\mathcal{M})$.

If $Ker\,\psi$ is also a two-sided ideal, Ψ is a linear map from the algebra $\mathcal{A}/Ker\,\psi$ to $C^\infty(\mathcal{M})$. Using this fact, we define a product, also denoted by $*$, on $\Psi(\mathcal{A}/Ker\,\psi) = \psi(\mathcal{A}) \subseteq C^\infty(\mathcal{M})$:

$$\Psi(\alpha + Ker\,\psi) * \Psi(\beta + Ker\,\psi) = \Psi\big((\alpha + Ker\,\psi)(\beta + Ker\,\psi)\big). \tag{3.10}$$

or

$$\psi(\alpha) * \psi(\beta) = \psi(\alpha\beta). \tag{3.11}$$

With this product, $\psi(\mathcal{A})$ is an algebra $(\psi(\mathcal{A}), *)$ isomorphic to $\mathcal{A}/Ker\,\psi$. (The notation means that $\psi(\mathcal{A})$ is considered with product $*$ and not say point-wise product).

We assume that $\mathcal{A}/Ker\,\psi$ is a $*$-algebra and that Ψ preserves the stars on $\mathcal{A}/Ker\,\psi$ and $C^\infty(\mathcal{M})$, the $*$ on the latter being complex conjugation denoted by bar:

$$\Psi\big((\alpha + Ker\,\psi)^*\big) = \overline{\Psi(\alpha + Ker\,\psi)}\,,$$
$$\psi(\alpha^*) = \overline{\psi(\alpha)}\,. \tag{3.12}$$

Such ψ and Ψ are said to be $*$-morphisms from $\mathcal{A}$ and $\mathcal{A}/Ker\,\psi$ to $(\psi(\mathcal{A}), *)$. The two algebras $\mathcal{A}/Ker\,\psi$ and $(\psi(\mathcal{A}), *)$ are $*$-isomorphic.

Remark: Star $(*)$ occurs with two meanings.

(1) It refers to involution on algebras in the phrase $*$-morphism.
(2) It refers to the new product on functions in $(\psi(\mathcal{A}), *)$.

These confusing notations, designed to keep the reader alert, are standard in the literature.

The above is the general framework. In applications, we encounter more than one linear bijection (one-to-one, onto map) from an a algebra $\mathcal{A}$ to $C^\infty(\mathcal{M})$ and that produces different-looking $*$'s on $C^\infty(\mathcal{M})$ and algebras $(C^\infty(\mathcal{M}), *)$, $(C^\infty(\mathcal{M}), *')$ etc. As they are $*$-isomorphic to $\mathcal{A}$, they are mutually $*$-isomorphic as well. A simple example we encounter below is $C^\infty(\mathbb{C})$ with Moyal- and coherent-state-induced $*$-products. These algebras are $*$-isomorphic.

3.2 Properties of Coherent States

It is useful to have the Campbell-Baker-Hausdorff (CBH) formula written down. It reads

$$e^{\hat{A}}e^{\hat{B}} = e^{\hat{A}+\hat{B}}e^{\frac{1}{2}[\hat{A},\hat{B}]} \tag{3.13}$$

for two operators $\hat{A}, \hat{B}$ if

$$[\hat{A}, [\hat{A}, \hat{B}]] = [\hat{B}, [\hat{A}, \hat{B}]] = 0\,. \tag{3.14}$$

For one oscillator with annihilation-creation operators $a, a^\dagger$, the coherent state

$$|z\rangle = e^{za^\dagger - \bar{z}a}|0\rangle = e^{-\frac{1}{2}|z|^2}e^{za^\dagger}|0\rangle\,, \quad z \in \mathbb{C} \tag{3.15}$$

has the properties

$$a|z\rangle = z|z\rangle\,; \qquad \langle z'|z\rangle = e^{\frac{1}{2}|z-z'|^2}\,. \qquad (3.16)$$

The coherent states are overcomplete, with the resolution of identity

$$\mathbf{1} = \int \frac{d^2 z}{\pi}|z\rangle\langle z|\,, \quad d^2 z = dx_1 dx_2\,, \quad \text{where} \quad z = \frac{x_1 + ix_2}{\sqrt{2}}\,. \qquad (3.17)$$

The factor $\frac{1}{\pi}$ is easily checked: $Tr\,\mathbf{1}|0\rangle\langle 0| = 1$ while $\int d^2 z|\langle 0|z\rangle|^2$ is π in view of (3.16).

A central property of coherent states is the following: an operator $\hat{A}$ is determined just by its diagonal matrix elements

$$A(z\,,\bar{z}) = \langle z|\hat{A}|z\rangle\,, \qquad (3.18)$$

that is by its "symbol" A, a function* on $\mathbb{C}$ with values $A(z\,,\bar{z}) = \langle z|\hat{A}|z\rangle$. An easy proof uses analyticity [54]. $\hat{A}$ is certainly determined by the collection of all its matrix elements $\langle \bar{\eta}|\hat{A}|\xi\rangle$ or equally by

$$e^{\frac{1}{2}(|\eta|^2 + |\xi|^2)}\langle \bar{\eta}|\hat{A}|\xi\rangle = \langle 0|e^{\eta a}\hat{A}e^{\xi a^\dagger}|0\rangle\,. \qquad (3.19)$$

The right hand side (at least for appropriate $\mathcal{A}$) is seen to be a holomorphic function of η and ξ, or equally well of

$$u = \frac{\eta + \xi}{2}\,, \quad v = \frac{\eta - \xi}{2i}\,. \qquad (3.20)$$

Holomorphic functions are globally determined by their values for real arguments. Hence the function $\tilde{A}$ defined by

$$\tilde{A}(u, v) = \langle 0|e^{\eta a^\dagger}\hat{A}e^{\xi a^\dagger}|0\rangle \qquad (3.21)$$

is globally determined by its values for u, v real or $\eta = \bar{\xi}$. Thus $\langle \xi|\hat{A}|\xi\rangle$ determines $\hat{A}$ as claimed.

There are also explicit formulas for $\hat{A}$ in terms of $\langle \xi|\bar{A}|\xi\rangle$ [55].

3.3 The Coherent State or Voros $*$-Product on the Moyal Plane

As indicated above, we can map an operator $\hat{A}$ to a function A using coherent states as follows:

$$\hat{A} \longrightarrow A\,, \quad A(z\,,\bar{z}) = \langle z|\hat{A}|z\rangle\,. \qquad (3.22)$$

---

*The $\bar{z}$ argument in $A(z\,,\bar{z})$ is redundant. It is there to emphasize that A is not necessarily a holomorphic function of the complex variable z.

This map is linear and also bijective by the previous remarks and induces a product $*_C$ on functions (C indicating "coherent state"). With this product, we get an algebra $(C^\infty(\mathbb{C}), *_C)$ of functions. Since the map $\hat{A} \to A$ has the property $\hat{A}^* \to A^* \equiv \bar{A}$, this map is a $*$-morphism from operators to $(C^\infty(\mathbb{C}), *_C)$.

Let us get familiar with this new function algebra.

The image of a is the function α where $\alpha(z, \bar{z}) = z$. The image of a^n has the value z^n at $(z, \bar{z})$, so by definition,

$$(\alpha *_C \alpha \ldots *_C \alpha)(z, \bar{z}) = z^n. \tag{3.23}$$

The image of $a^* \equiv a^\dagger$ is $\bar{\alpha}$ where $\bar{\alpha}(z, \bar{z}) = \bar{z}$ and that of $(a^*)^n$ is $\bar{\alpha} *_C \bar{\alpha} \cdots *_C \bar{\alpha}$ where

$$(\bar{\alpha} *_C \bar{\alpha} \cdots *_C \bar{\alpha})(z, \bar{z}) = \bar{z}^n. \tag{3.24}$$

Since $\langle z | a^* a | z \rangle = \bar{z} z$ and $\langle z | a a^* | z \rangle = \bar{z} z + 1$, we get

$$\bar{\alpha} *_C \alpha = \bar{\alpha} \alpha, \qquad \alpha *_C \bar{\alpha} = \alpha \bar{\alpha} + \mathbf{1}, \tag{3.25}$$

where $\bar{\alpha} \alpha = \alpha \bar{\alpha}$ is the pointwise product of α and $\bar{\alpha}$, and $\mathbf{1}$ is the constant function with value 1 for all z.

For general operators $\hat{f}$, the construction proceeds as follows. Consider

$$: e^{\xi a^\dagger - \bar{\xi} a} : \tag{3.26}$$

where the normal ordering symbol $: \cdots :$ means as usual that $a^\dagger$'s are to be put to the left of a's. Thus

$$: a a^\dagger a^\dagger a : = a^\dagger a^\dagger a a,$$
$$: e^{\xi a^\dagger - \bar{\xi} a} : = e^{\xi a^\dagger} e^{-\bar{\xi} a}. \tag{3.27}$$

Hence

$$\langle z | : e^{\xi a^\dagger - \bar{\xi} a} : | z \rangle = e^{\xi \bar{z} - \bar{\xi} z}. \tag{3.28}$$

Writing $\hat{f}$ as a Fourier transform,

$$\hat{f} = \int \frac{d^2 \xi}{\pi} : e^{\xi a^\dagger - \bar{\xi} a} : \tilde{f}(\xi, \bar{\xi}), \qquad \tilde{f}(\xi, \bar{\xi}) \in \mathbb{C}, \tag{3.29}$$

its symbol is seen to be

$$f = \int \frac{d^2 \xi}{\pi} e^{\xi \bar{z} - \bar{\xi} z} \tilde{f}(\xi, \bar{\xi}). \tag{3.30}$$

This map is invertible since f determines $\tilde{f}$.

Consider also the second operator

$$\hat{g} = \int \frac{d^2\eta}{\pi} \, : e^{\eta a^\dagger - \bar{\eta} a} : \tilde{g}(\eta,\bar{\eta}) \,, \qquad (3.31)$$

and its symbol

$$g = \int \frac{d^2\eta}{\pi} e^{\eta\bar{z} - \bar{\eta}z} \tilde{g}(\eta,\bar{\eta}) \,. \qquad (3.32)$$

The task is to find the symbol $f *_C g$ of $\hat{f}\hat{g}$.

Let us first find

$$e^{\xi\bar{z} - \bar{\xi}z} *_C e^{\eta\bar{z} - \bar{\eta}z} \,. \qquad (3.33)$$

We have

$$: e^{\xi a^\dagger - \bar{\xi}a} :\, : e^{\eta a^\dagger - \bar{\eta}a} :=: e^{\xi a^\dagger - \bar{\xi}a} \, e^{\eta a^\dagger - \bar{\eta}a} : e^{-\bar{\xi}\eta} \qquad (3.34)$$

and hence

$$e^{\xi\bar{z} - \bar{\xi}z} *_C e^{\eta\bar{z} - \bar{\eta}z} = e^{-\bar{\xi}\eta} e^{\xi\bar{z} - \bar{\xi}z} \, e^{\eta\bar{z} - \bar{\eta}z}$$

$$= e^{\xi\bar{z} - \bar{\xi}z} e^{\overleftarrow{\partial}_z \overrightarrow{\partial}_{\bar{z}}} e^{\eta\bar{z} - \bar{\eta}z} \,. \qquad (3.35)$$

The bidifferential operators $\left(\overleftarrow{\partial}_z \overrightarrow{\partial}_{\bar{z}}\right)^k$, $(k = 1, 2, \ldots)$ have the definition

$$\alpha\left(\overleftarrow{\partial}_z \overrightarrow{\partial}_{\bar{z}}\right)^k \beta\,(z,\bar{z}) = \frac{\partial^k \alpha(z,\bar{z})}{\partial z^k} \frac{\partial^k \beta(z,\bar{z})}{\partial \bar{z}^k} \,. \qquad (3.36)$$

The exponential in (3.35) involving them can be defined using the power series.

$f *_C g$ follows from (3.35):

$$(f *_C g)(z,\bar{z}) = \left(f e^{\overleftarrow{\partial}_z \overrightarrow{\partial}_{\bar{z}}} g\right)(z,\bar{z}) \,. \qquad (3.37)$$

(3.37) is the coherent state $*$-product [56].

We can explicitly introduce a deformation parameter $\theta > 0$ in the discussion by changing (3.37) to

$$f *_C g(z,\bar{z}) = \left(f e^{\theta \overleftarrow{\partial}_z \overrightarrow{\partial}_{\bar{z}}} g\right)(z,\bar{z}) \,. \qquad (3.38)$$

After rescaling $z' = \frac{z}{\sqrt{\theta}}$, (3.38) gives (3.37). As z' and $\bar{z}'$ after quantization become $a, a^\dagger$, z and $\bar{z}$ become the scaled oscillators $a_\theta, a_\theta^\dagger$:

$$[a_\theta, a_\theta] = [a_\theta^\dagger, a_\theta^\dagger] = 0 \,, \quad [a_\theta, a_\theta^\dagger] = \theta \,. \qquad (3.39)$$

(3.39) is associated with the Moyal plane with Cartesian coordinate functions x_1, x_2. If $a_\theta = \frac{x_1 + ix_2}{\sqrt{2}}, a_\theta^\dagger = \frac{x_1 - ix_2}{\sqrt{2}}$,

$$[x_i, x_j] = i\theta\varepsilon_{ij} \,, \quad \varepsilon_{ij} = -\varepsilon_{ji} \,, \quad \varepsilon_{12} = 1 \,. \qquad (3.40)$$

The Moyal plane is the plane $\mathbb{R}^2$, but with its function algebra deformed in accordance with (3.40). The deformed algebra has the product (3.38) or equivalently the Moyal product derived below.

Discussion of the equivalence between Moyal and Voros star products, using Stratonovich's methods [57] and with several physical applications is given in [58, 59].

3.4 The Moyal ∗-Product on the Groenewold-Moyal Plane

We get this by changing the map $\hat{f} \to f$ from operators to functions. For a given function f, the operator $\hat{f}$ is thus different for the coherent state and Moyal ∗'s. The ∗-product on two functions is accordingly also different.

3.4.1 The Weyl Map and the Weyl Symbol

The Weyl map of the operator

$$\hat{f} = \int \frac{d^2\xi}{\pi} \tilde{f}(\xi,\bar{\xi}) e^{\xi a^\dagger - \bar{\xi} a} , \tag{3.41}$$

to the function f is defined by

$$f(z,\bar{z}) = \int \frac{d^2\xi}{\pi} \tilde{f}(\xi,\bar{\xi}) e^{\xi\bar{z} - \bar{\xi} z} . \tag{3.42}$$

(3.42) makes sense since $\tilde{f}$ is fully determined by $\hat{f}$ as follows:

$$\langle z|\hat{f}|z\rangle = \int \frac{d^2\xi}{\pi} \tilde{f}(\xi,\bar{\xi}) e^{-\frac{1}{2}\xi\bar{\xi}} e^{\xi\bar{z} - \bar{\xi} z} . \tag{3.43}$$

$\tilde{f}$ can be calculated from here by Fourier transformation.

The map is invertible since $\tilde{f}$ follows from f by Fourier transform of (3.42) and $\tilde{f}$ fixes $\hat{f}$ by (3.41).

f is called the *Weyl symbol* of $\hat{f}$.

As the Weyl map is bijective, we can find a new ∗ product, call it $*_W$, between functions by setting $f *_W g = $ Weyl symbol of $\hat{f}\hat{g}$.

For

$$\hat{f} = e^{\xi a^\dagger - \bar{\xi} a} , \quad \hat{g} = e^{\eta a^\dagger - \bar{\eta} a} , \tag{3.44}$$

to find $f *_W g$, we first rewrite $\hat{f}\hat{g}$ according to

$$\hat{f}\hat{g} = e^{\frac{1}{2}(\xi\bar{\eta} - \bar{\xi}\eta)} e^{(\xi+\eta)a^\dagger - (\bar{\xi}+\bar{\eta})a} . \tag{3.45}$$

Hence

$$\begin{aligned}
(f *_W g)(z,\bar{z}) &= e^{\xi\bar{z} - \bar{\xi} z} e^{\frac{1}{2}(\xi\bar{\eta} - \bar{\xi}\eta)} e^{\eta\bar{z} - \bar{\eta} z} \\
&= f e^{\frac{1}{2}\left(\overleftarrow{\partial}_z \overrightarrow{\partial}_{\bar{z}} - \overleftarrow{\partial}_{\bar{z}} \overrightarrow{\partial}_z\right)} g(z,\bar{z}) .
\end{aligned} \tag{3.46}$$

Multiplying by $\tilde{f}(\xi,\bar{\xi})$, $\tilde{g}(\eta,\bar{\eta})$ and integrating, we get (3.46) for arbitrary functions:

$$(f *_W g)(z,\bar{z}) = \left(f e^{\frac{1}{2}\left(\overleftarrow{\partial}_z \overrightarrow{\partial}_{\bar{z}} - \overleftarrow{\partial}_{\bar{z}} \overrightarrow{\partial}_z\right)} g\right)(z,\bar{z}) . \tag{3.47}$$

Note that

$$\overleftarrow{\partial}_z \overrightarrow{\partial}_{\bar z} - \overleftarrow{\partial}_{\bar z} \overrightarrow{\partial}_z = i(\overleftarrow{\partial}_1 \overrightarrow{\partial}_2 - \overleftarrow{\partial}_2 \overrightarrow{\partial}_1) = i\varepsilon_{ij} \overleftarrow{\partial}_i \overrightarrow{\partial}_j . \qquad (3.48)$$

Introducing also θ, we can write the $*_W$-product as

$$f *_W g = f e^{i\frac{\theta}{2}\varepsilon_{ij}\overleftarrow{\partial}_i \overrightarrow{\partial}_j} g . \qquad (3.49)$$

By (3.40), $\theta\varepsilon_{ij} = \omega_{ij}$ fixes the Poisson brackets, or the Poisson structure on the Moyal plane. (3.49) is customarily written as

$$f *_W g = f e^{\frac{i}{2}\omega_{ij}\overleftarrow{\partial}_i \overrightarrow{\partial}_j} g . \qquad (3.50)$$

using the Poisson structure. (But we have not cared to position the indices so as to indicate their tensor nature and to write ω^{ij}.)

3.5 Properties of $*$-Products

A $*$-product without a subscript indicates that it can be either a $*_C$ or a $*_W$.

3.5.1 *Cyclic Invariance*

The trace of operators has the fundamental property

$$Tr \hat{A}\hat{B} = Tr \hat{B}\hat{A} \qquad (3.51)$$

which leads to the general cyclic identities

$$Tr \hat{A}_1 \ldots \hat{A}_n = Tr \hat{A}_n \hat{A}_1 \ldots \hat{A}_{n-1} . \qquad (3.52)$$

We now show that

$$Tr \hat{A}\hat{B} = \int \frac{d^2 z}{\pi} A * B(z, \bar z), \qquad * = *_C \quad \text{or} \quad *_W . \qquad (3.53)$$

(The functions on R.H.S. are different for $*_C$ and $*_W$ if $\hat{A}, \hat{B}$ are fixed). From this follows the analogue of (3.52):

$$\int \frac{d^2 z}{\pi} (A_1 * A_2 * \cdots * A_n)(z, \bar z) = \int \frac{d^2 z}{\pi} (A_n * A_1 * \cdots * A_{n-1})(z, \bar z) . \qquad (3.54)$$

For $*_C$, (3.53) follows from (3.17).

The coherent state image of $e^{\xi a^\dagger - \bar\xi a}$ is the function with value

$$e^{\xi\bar z - \bar\xi z} e^{-\frac{1}{2}\bar\xi\xi} \qquad (3.55)$$

at z, with a similar correspondence if $\xi \to \eta$. So

$$Tr\, e^{\xi a^\dagger - \bar{\xi}a}\, e^{\eta a^\dagger - \bar{\eta}a} = \int \frac{d^2z}{\pi} \left(e^{\xi\bar{z}-\bar{\xi}z}e^{-\frac{1}{2}\bar{\xi}\xi} \right) \left(e^{\eta\bar{z}-\bar{\eta}z}e^{-\frac{1}{2}\bar{\eta}\eta} \right) e^{-\bar{\xi}\eta} \quad (3.56)$$

The integral produces the δ-function

$$\prod_i 2\delta(\xi_i + \eta_i)\,, \qquad \xi_i = \frac{\xi_1 + \xi_2}{\sqrt{2}}\,, \qquad \eta_i = \frac{\eta_1 + \eta_2}{\sqrt{2}}\,. \quad (3.57)$$

We can hence substitute $e^{-\left(\frac{1}{2}\bar{\xi}\xi+\frac{1}{2}\bar{\eta}\eta+\bar{\xi}\eta\right)}$ by $e^{\frac{1}{2}(\xi\bar{\eta}-\bar{\xi}\eta)}$ and get (3.53) for Weyl $*$ for these exponentials and so for general functions by using (3.41).

3.5.2 A Special Identity for the Weyl Star

The above calculation also gives, the identity

$$\int \frac{d^2z}{\pi}(A *_W B)(z\,,\bar{z}) = \int \frac{d^2z}{\pi} A(z\,,\bar{z})\, B(z\,,\bar{z})\,. \quad (3.58)$$

That is because

$$\prod_i \delta(\xi_i + \eta_i)\, e^{\frac{1}{2}(\xi\bar{\eta}-\bar{\xi}\eta)} = \prod_i \delta(\xi_i + \eta_i)\,. \quad (3.59)$$

In (3.54), A and B in turn can be Weyl $*$-products of other functions. Thus in integrals of Weyl $*$-products of functions, one $*_W$ can be replaced by the pointwise (commutative) product:

$$\int \frac{d^2z}{\pi} \left(A_1 *_W A_2 *_W \cdots A_K \right) *_W \left(B_1 *_W B_2 *_W \cdots B_L \right)(z\,,\bar{z})$$

$$= \int \frac{d^2z}{\pi} \left(A_1 *_W A_2 *_W \cdots A_K \right) \left(B_1 *_W B_2 *_W \cdots B_L \right)(z\,,\bar{z})\,. \quad (3.60)$$

This identity is frequently useful.

3.5.3 Equivalence of $*_C$ and $*_W$

For the operator

$$\hat{A} = e^{\xi a^\dagger - \bar{\xi}a}\,, \quad (3.61)$$

the coherent state function A_C has the value (3.55) at z, and the Weyl symbol A_W has the value

$$A_W(z\,,\bar{z}) = e^{\xi\bar{z}-\bar{\xi}z}\,. \quad (3.62)$$

As both $\left(C^\infty(\mathbb{R}^2), *_C\right)$ and $\left(C^\infty(\mathbb{R}^2), *_W\right)$ are isomorphic to the operator algebra, they too are isomorphic. The isomorphism is established by the maps

$$A_C \longleftrightarrow A_W \tag{3.63}$$

and their extension via Fourier transform to all operators $\hat{A}$ and functions $A_{C,W}$.

Clearly

$$A_W = e^{-\frac{1}{2}\partial_z \partial_{\bar{z}}} A_C, \quad A_C = e^{\frac{1}{2}\partial_z \partial_{\bar{z}}} A_W,$$

$$A_C *_C B_C \longleftrightarrow A_W *_W B_W. \tag{3.64}$$

The mutual isomorphism of these three algebras is a $*$-isomorphism since $(\hat{A}\hat{B})^\dagger \longrightarrow \bar{B}_{C,W} *_{C,W} \bar{A}_{C,W}$.

3.5.4 *Integration and Tracial States*

This is a good point to introduce the ideas of a state and a tracial state on a $*$-algebra $\mathcal{A}$ with unity $\mathbf{1}$.

A state ω is a linear map from $\mathcal{A}$ to $\mathbb{C}$, $\omega(a) \in \mathbb{C}$ for all $a \in \mathcal{A}$ with the following properties:

$$\omega(a^*) = \overline{\omega(a)},$$
$$\omega(a^*a) \geq 0,$$
$$\omega(\mathbf{1}) = 1. \tag{3.65}$$

If $\mathcal{A}$ consists of operators on a Hilbert space and ρ is a density matrix, it defines a state ω_ρ via

$$\omega_\rho(a) = Tr(\rho a). \tag{3.66}$$

If $\rho = e^{-\beta H}/Tr(e^{-\beta H})$ for a Hamiltonian H, it gives a Gibbs state via (3.66).

Thus the concept of a state on an algebra $\mathcal{A}$ generalizes the notion of a density matrix. There is a remarkable construction, the Gel'fand- Naimark-Segal (GNS) construction which shows how to associate any state with a rank-1 density matrix [60].

A state is *tracial* if it has cyclic invariance [60]:

$$\omega(ab) = \omega(ba). \tag{3.67}$$

The Gibbs state is not tracial, but fulfills an identity generalizing (3.67). It is a Kubo-Martin-Schwinger (KMS) state [60].

A positive map ω' is in general an unnormalized state: It must fulfill all the conditions that a state fulfills, but is not obliged to fulfill the condition $\omega'(1) = 1$.

Let us define a positive map ω' on $(C^\infty(\mathbb{R}^2), *)$ ($* = *_C$ or $*_W$) using integration:

$$\omega'(A) = \int \frac{d^2 z}{\pi} \, \hat{A}(z, \bar{z}) \,. \tag{3.68}$$

It is easy to verify that ω' fulfills the properties of a positive map.

A *tracial* positive map ω' also has the cyclic invariance (3.67).

The cyclic invariance (3.67) of $\omega'(A*B)$ means that it is a tracial positive map.

3.5.5 *The θ-Expansion*

On introducing θ, we have (3.38) and

$$f *_W g(z, \bar{z}) = f e^{\frac{\theta}{2} \left(\overleftarrow{\partial}_z \overrightarrow{\partial}_{\bar{z}} - \overleftarrow{\partial}_{\bar{z}} \overrightarrow{\partial}_z \right)} g(z, \bar{z}) \,. \tag{3.69}$$

The series expansion in θ is thus

$$f *_C g(z, \bar{z}) = f g(z, \bar{z}) + \theta \frac{\partial f}{\partial z}(z, \bar{z}) \frac{\partial g}{\partial \bar{z}}(z, \bar{z}) + \mathcal{O}(\theta^2) \,, \tag{3.70}$$

$$f *_W g(z, \bar{z}) = f g(z, \bar{z}) + \frac{\theta}{2} \left(\frac{\partial f}{\partial z} \frac{\partial g}{\partial \bar{z}} - \frac{\partial f}{\partial \bar{z}} \frac{\partial g}{\partial z} \right)(z, \bar{z}) + \mathcal{O}(\theta^2) \,. \tag{3.71}$$

Introducing the notation

$$[f, g]_* = f * g - g * f \,, \quad * = *_C \quad \text{or} \quad *_W \,, \tag{3.72}$$

We see that

$$[f, g]_{*_C} = \theta \left(\frac{\partial f}{\partial z} \frac{\partial g}{\partial \bar{z}} - \frac{\partial f}{\partial \bar{z}} \frac{\partial g}{\partial z} \right)(z, \bar{z}) + \mathcal{O}(\theta^2) \,,$$

$$[f, g]_{*_W} = \theta \left(\frac{\partial f}{\partial z} \frac{\partial g}{\partial \bar{z}} - \frac{\partial f}{\partial \bar{z}} \frac{\partial g}{\partial z} \right)(z, \bar{z}) + \mathcal{O}(\theta^2) \,, \tag{3.73}$$

We thus see that

$$[f, g]_* = i\theta \{f, g\}_{P.B.} + \mathcal{O}(\theta^2) \,, \tag{3.74}$$

where $\{f, g\}$ is the Poisson Bracket of f and g and the $\mathcal{O}(\theta^2)$ term depends on $*_{C,W}$. Thus the $*$-product is an associative product which to leading order in the deformation parameter ("Planck's" constant) θ is compatible with the rules of quantization of Dirac. We can say that with the $*$-product, we have deformation quantization of the classical commutative algebra of functions.

But it should be emphasized that even to leading order in θ, $f *_C g$ and $f *_W g$ do not agree. Still the algebras $\left(C^\infty(\mathbb{R}^2, *_C)\right)$ and $\left(C^\infty(\mathbb{R}^2, *_W)\right)$ are $*$-isomorphic.

Suppose we are given a Poisson structure on a manifold M with Poisson bracket $\{.\,,.\}$. Then Kontsevich ([61]) has given a $*$-product $f * g$ as a formal power series in θ such that (3.74) holds.

For further developments on the physical equivalence of $*_C$ and $*_W$ see, [62].

3.6 The $*$-Products for the Fuzzy Sphere

Star products for Kähler manifolds have been known for a long time. The approach we take here was initiated by Prešnajder [63](See also [62], it produces particularly compact expressions.).

Let P_n be the orthogonal projection operator to the subspace with $N = n$. The fuzzy sphere algebra is then the algebra with elements $P_n \gamma(a_i^\dagger a_j) P_n$ where γ is any polynomial in $a_i^\dagger a_j$. As any such polynomial commutes with N, if γ and δ are two of these polynomials,

$$P_n \gamma(a_i^\dagger a_j) P_n P_n \delta(a_i^\dagger a_j) P_n = P_n \gamma(a_i^\dagger a_j) \delta(a_i^\dagger a_j) P_n \,. \tag{3.75}$$

This algebra, more precisely, is the orthogonal direct sum $Mat(n+1) \oplus 0$ where $Mat(n+1)$ acts on the $\widehat{N} = n$ subspace and is the fuzzy sphere. But the extra 0 here is entirely harmless.

3.6.1 *The Coherent State $*$-Product $*_C$*

There are now two oscillators a_1, a_2, so the coherent states are labeled by two complex variables, being

$$|Z_1, Z_2\rangle = e^{Za^\dagger - \bar{Z}a}|0\rangle, \qquad Z = (Z_1, Z_2)\,. \tag{3.76}$$

We use capital Z's for unnormalized Z's and z's for normalized ones: $z = \frac{Z}{|Z|}$, $|Z|^2 = \sum |Z_i|^2$.

The normalized coherent states $|z\rangle_n$ for S_F^2, as one can guess, are obtained by projection from $|Z\rangle$,

$$|z\rangle_n = \frac{P_n|Z\rangle}{|\langle P_n|Z\rangle|} = \frac{\left(\sum_i z_i a_i^\dagger\right)^n}{\sqrt{n!}}|0\rangle\,, \tag{3.77}$$

where we have used

$$P_n|Z\rangle = \frac{(Z_i a_i^\dagger)^n}{n!}|0\rangle. \tag{3.78}$$

They are called Perelomov coherent states [54]

For an operator $P_n \hat{A} P_n$, the coherent state symbol has the value

$$\langle Z|P_n \hat{A} P_n|Z\rangle = \frac{|z|^{2n}}{n!}\langle z|\hat{A}|z\rangle_n \tag{3.79}$$

at Z. By a previous result, the diagonal coherent state expectation values $\langle z|P_n \hat{A} P_n|z\rangle_n$ determines $P_n \hat{A} P_n$ uniquely and there is a $*$-product for S_F^2. We call it a $*_C$-product in analogy to the notation used before.

We can find it explicitly as follows [63, 31, 9]. For $n = 1$ (spin $\frac{n}{2} = \frac{1}{2}$), a basis for 2×2 matrices is

$$\{\sigma_A : \sigma_0 = \mathbf{1}, \sigma_i \quad (i = 1, 2, 3) \quad = \quad \text{Pauli Matrices}, \quad Tr\sigma_A\sigma_B = 2\delta_{AB}\}. \tag{3.80}$$

Let

$$|i\rangle = a_i^\dagger|0\rangle, \quad i = 1, 2 \tag{3.81}$$

be a pair of orthonormal vectors for $n = 1$. A general operator is

$$\hat{F} = f_A\hat{\sigma}_A, \quad \hat{\sigma}_A = a^\dagger\sigma_A a\big|_{n=1}, \quad f_A \in \mathbb{C}. \tag{3.82}$$

and $\hat{\sigma}_A|i\rangle = |j\rangle(\sigma_A)_{ji}$. In above, by $a^\dagger\sigma_A a\big|_{n=1}$, we mean the restriction of $a^\dagger\sigma_A a$ to the subspace with $n = 1$.

Call the coherent state symbol of $\hat{\sigma}_A$ for $n = 1$ as χ_A:

$$\chi_A(z) = \langle z|\hat{\sigma}_A|z\rangle, \quad \chi_0(z) = 1, \quad \chi_i = \bar{z}\sigma_i z, \quad i = 1, 2, 3. \tag{3.83}$$

The $*$-product for $n = 1$ now follows:

$$\chi_A *_C \chi_B(z) = \langle z|\hat{\sigma}_A\hat{\sigma}_B|z\rangle. \tag{3.84}$$

Write

$$\sigma_A\sigma_B = \delta_{AB} + E_{ABi}\sigma_i \tag{3.85}$$

to get

$$\chi_A *_C \chi_B(z) = \delta_{AB} + E_{ABi}\chi_i(z)$$
$$:= \chi_A(z)\chi_B(z) + \mathcal{K}_{AB}(z). \tag{3.86}$$

Let us use the notation

$$n_i = \chi_i(z), \quad n_0 = 1. \tag{3.87}$$

$\vec{n}$ is the coordinate on S^2: $\vec{n} \cdot \vec{n} = 1$. Then (3.86) is

$$n_A *_C n_B(z) = n_A n_B + K_{AB}(n), \qquad \mathcal{K}_{AB}(z) := K_{AB}(n). \qquad (3.88)$$

This K_{AB} has a particular significance for complex analysis. Since $\chi_0(z) = 1$, $\chi_0(z) * \chi_A = \chi_0 \chi_A$ by (3.86) and

$$K_{0A} = 0. \qquad (3.89)$$

The components $K_{ij}(n)$ of K can be calculated from (3.85), (3.86). Let $\theta(\alpha)$ be the spin 1 angular momentum matrices:

$$\theta(\alpha)_{ij} = -i\varepsilon_{\alpha ij}. \qquad (3.90)$$

Then

$$K_{ij}(\vec{n}) = \frac{\{\vec{\theta} \cdot \vec{n} (\vec{\theta} \cdot \vec{n} - 1)\}_{ij}}{2}$$

$$\vec{\theta} \cdot \vec{n} := \theta(\alpha)n_\alpha. \qquad (3.91)$$

The eigenvalues of $\vec{\theta} \cdot \vec{n}$ are $\pm 1, 0$ and $K_{ij}(\vec{n})$ is the projection operator to the eigenspace $\vec{\theta} \cdot \vec{n} = -1$,

$$K(\vec{n})^2 = K(\vec{n}). \qquad (3.92)$$

It is related to the complex structure of S^2 in the projective module picture treated in chapter 5.

The vector space for angular momentum $\frac{n}{2}$ is the n-fold symmetric tensor product of the spin-$\frac{1}{2}$ vector spaces. The general linear operator on this space can be written as

$$\widehat{F} = f_{A_1 A_2 \cdots A_n} \hat{\sigma}_{A_1} \otimes \hat{\sigma}_{A_2} \otimes \cdots \hat{\sigma}_{A_n} \qquad (3.93)$$

where f is totally symmetric in its indices. Its symbol is thus

$$F(\vec{n}) = f_{A_1 A_2 \cdots A_n} n_{A_1} n_{A_2} \cdots n_{A_n}, \qquad n_0 := 1. \qquad (3.94)$$

The symbol of another operator

$$\widehat{G} = g_{B_1 B_2 \cdots B_n} \hat{\sigma}_{B_1} \otimes \hat{\sigma}_{B_2} \otimes \cdots \hat{\sigma}_{B_n}, \qquad (3.95)$$

where g is symmetric in its indices, is

$$G(\vec{n}) = g_{B_1 B_2 \cdots B_n} n_{B_1} n_{B_2} \cdots n_{B_n}. \qquad (3.96)$$

Since

$$\widehat{F}\widehat{G} = f_{A_1 A_2 \cdots A_n} g_{B_1 B_2 \cdots B_n} \sigma_{A_1} \sigma_{B_1} \otimes \sigma_{A_2} \sigma_{B_2} \otimes \cdots \otimes \sigma_{A_n} \sigma_{B_n}, \qquad (3.97)$$

we have that

$$F * G(\vec{n}) = f_{A_1 A_2 \cdots A_n} g_{B_1 B_2 \cdots B_n} \prod_i \left(n_{A_i} n_{B_i} + K_{A_i B_i} \right) \qquad (3.98)$$

or

$$F * G(\vec{n}) =$$

$$FG(\vec{n}) + \sum_{m=1}^{n} \frac{n!}{m!(n-m)!} f_{A_1 A_2 \cdots A_m A_{m+1} \cdots A_n} n_{A_{m+1}} n_{A_{m+2}} \cdots n_{A_n}$$

$$\times K_{A_1 B_1}(\vec{n}) K_{A_2 B_2}(\vec{n}) \cdots K_{A_m B_m}(\vec{n}) g_{B_1 B_2 \cdots B_m B_{m+1} \cdots B_n} n_{B_{m+1}} \times$$

$$n_{B_{m+2}} \cdots n_{B_n}. \tag{3.99}$$

Now as f and g are symmetric in indices, there is the expression

$$\partial_{A_1} \partial_{A_2} \cdots \partial_{A_m} F(\vec{n}) = \frac{n!}{(n-m)!} f_{A_1 A_2 \cdots A_m A_{m+1} \cdots A_n} n_{A_{m+1}} n_{A_{m+2}} \cdots n_{A_n},$$

$$\partial_{A_i} \equiv \frac{\partial}{\partial n_{A A_i}} \tag{3.100}$$

for F and a similar expression for G. Hence

$$F *_C G(\vec{n}) = \sum_{m=0}^{n} \frac{(n-m)!}{m! n!} \left(\partial_{A_1} \partial_{A_2} \cdots \partial_{A_m} F \right)(\vec{n})$$

$$\times K_{A_1 B_1}(\vec{n}) K_{A_2 B_2}(\vec{n}) \cdots K_{A_m B_m}(\vec{n}) \left(\partial_{B_1} \partial_{B_2} \cdots \partial_{B_m} G \right)(\vec{n}) \tag{3.101}$$

which is the final answer. Here the $m = 0$ term is to be understood as $FG(\vec{n})$, the pointwise product of F and G evaluated at $\vec{n}$. This formula was first given in [63].

Differentiating on n_A ignoring the constraint $\vec{n} \cdot \vec{n} = 1$ is justified in the final answer (3.101) (although not in (3.100)), since $K_{AB}(\vec{n}) \partial_A (\vec{n} \cdot \vec{n}) = K_{AB}(\vec{n}) \partial_B (\vec{n} \cdot \vec{n}) = 0$. (3.100) being only an intermediate step on the way to (3.101), this sloppiness is immaterial.

For large n, (3.101) is an expansion in powers of $\frac{1}{n}$, the leading term giving the commutative product. Thus the algebra S_F^2 is in some sense a deformation of the commutative algebra of functions $C^\infty(S^2)$. But as the maximum angular momentum in F and G is n, we get only the spherical harmonics $Y_{\ell m}, \ell \in \{0, 1, \cdots n\}$ in their expansion. For this reason, F and G span a finite-dimensional subspace of $C^\infty(S^2)$ and S_F^2 is not properly a deformation of the commutative algebra $C^\infty(S^2)$.

3.6.2 The Weyl *-Product $*_W$

The Weyl *-products are characterized by the special identity described before. For this reason they are very convenient for use in loop expansions in quantum field theory (see chapter 4).

A simple way to find $*_W$ is to find it via its connection to $*_C$. For this purpose let us consider

$$Tr(\hat{T}_m^\ell)^\dagger \hat{T}_{m'}^{\ell'} = \frac{n+1}{4\pi} \int d\Omega [T_n(\ell)^{\frac{1}{2}} \overline{Y}_{\ell m}] *_C [T_n(\ell')^{\frac{1}{2}} Y_{\ell'm'}](\vec{x}), \quad (3.102)$$

where

$$\langle z, n | T_m^\ell | z, n \rangle = T_n(\ell)^{\frac{1}{2}} Y_{\ell m}(\hat{n}). \quad (3.103)$$

T_m^ℓ is defined in section 2.5 while we can show (3.103) as follows. The factor $T_n(\ell)^{\frac{1}{2}}$ is independent of m by rotational invariance. It is real as shown by complex conjugating (3.103) and using

$$(T_m^\ell)^\dagger = (-1)^m T_{-m}^\ell, \quad \overline{Y}_{\ell m}(\hat{n}) = (-1)^m Y_{\ell,-m}(\vec{n}). \quad (3.104)$$

It can be chosen to be positive as well by explicit calculation. We shall evaluate it later.

The normalization of T_m^ℓ and $Y_{\ell m}$ are

$$Tr(T_m^\ell)^\dagger T_{m'}^{\ell'} = \int d\Omega \overline{Y}_{\ell m}(\vec{x}) Y_{\ell'm'}(\vec{x}) = \delta_{\ell\ell'} \delta_{mm'}. \quad (3.105)$$

Hence using (3.102)

$$\begin{aligned}
\delta_{\ell\ell'} \delta_{mm'} &= \int d\Omega \overline{Y}_{\ell m}(\vec{x}) Y_{\ell'm'}(\vec{x}) \\
&= \frac{n+1}{4\pi} \int d\Omega (T_n(\ell)^{\frac{1}{2}} \overline{Y}_{\ell m})(\vec{x}) *_C (T_n(\ell')^{\frac{1}{2}} Y_{\ell'm'})(\vec{x}) \quad (3.106)
\end{aligned}$$

Equation (3.106) suggests that the fuzzy sphere algebra $(S_F^2, *_W)$ with the Weyl-Moyal product $*_M$ is obtained from the fuzzy sphere algebra $(S_F^2, *_C)$ with the coherent state $*_C$ product from the map

$$\chi : (S_F^2, *_C) \longrightarrow (S_F^2, *_W)$$

$$\chi\left(\sqrt{\frac{n+1}{4\pi}} T_n(\ell)^{\frac{1}{2}} Y_{\ell m}\right) = Y_{\ell m}. \quad (3.107)$$

The induced $*$, call it for a moment as $*'$, on the image of χ is

$$Y_{\ell m} *' Y_{\ell'm'} = \chi\left(\sqrt{\frac{n+1}{4\pi}} T_n(\ell)^{\frac{1}{2}} Y_{\ell m} *_C \sqrt{\frac{n+1}{4\pi}} T_n(\ell')^{\frac{1}{2}} Y_{\ell'm'}\right). \quad (3.108)$$

For the evaluation of (3.108), $Y_{\ell m} *_C Y_{\ell'm'}$ has to be written as a series in $Y_{\ell''m''}$ and χ applied to it term-by-term. We will not need its full details here.

Now replace $Y_{\ell m}$ by $\overline{Y}_{\ell m}$ and integrate. As χ commutes with rotations, only the angular momentum 0 component of $\sqrt{\frac{n+1}{4\pi}} T_n(\ell)^{\frac{1}{2}} \overline{Y}_{\ell m} *_C$

$\sqrt{\frac{n+1}{4\pi}}T_n(\ell')^{\frac{1}{2}}Y_{\ell'm'}$ contributes to the integral. This component is $\delta_{\ell\ell'}\delta_{mm'}\overline{Y}_{00}*_C Y_{00} = \delta_{\ell\ell'}\delta_{mm'}\frac{1}{4\pi}$. Using (3.107) for $\ell = 0$ and the value $T_n(0)^{\frac{1}{2}} = \sqrt{\frac{4\pi}{n+1}}$ to be derived below, we get

$$\int d\Omega \overline{Y}_{\ell m}*' Y_{\ell'm'} = \delta_{\ell\ell'}\delta_{mm'} = \int d\Omega \overline{Y}_{\ell m}Y_{\ell'm'}. \tag{3.109}$$

Hence $*'$ enjoys the special identity characterizing the Weyl-Moyal product for the basis of functions in our algebra and hence for all functions. $*'$ is the Weyl-Moyal product $*_W$.

T_n is a function $\mathcal{T}_n$ of $\ell(\ell+1)$. The latter is the eigenvalue of $\mathcal{L}^2$, the square of angular momentum. The map χ can hence be defined directly on all functions α by

$$\chi(\alpha) = \sqrt{\frac{n+1}{4\pi}}\mathcal{T}_n(\mathcal{L}^2)^{\frac{1}{2}}\alpha \tag{3.110}$$

where R.H.S. can be calculated for example by expanding α in spherical harmonics.

The evaluation of $T_n^{\frac{1}{2}}(\ell)$ can be done as follows. It is enough to compare the two sides of (3.103) for $m = \ell$. For $m = \ell$,

$$Y_{\ell\ell}(\vec{x}) = \frac{\sqrt{(2\ell+1)!}}{\ell!}\bar{z}_2^{\ell}z_1^{\ell}. \tag{3.111}$$

The operator T_ℓ^ℓ being the highest weight state commutes with $L_+ = a_2^\dagger a_1$ while $[L_3, T_\ell^\ell] = \ell T_\ell^\ell$. Hence in terms of a_i and $a_j^\dagger$,

$$T_\ell^\ell = N_\ell a_2^{\dagger\ell}a_1^\ell \tag{3.112}$$

where the constant N_ℓ is to be fixed by the condition

$$Tr(T_\ell^\ell)^\dagger T_\ell^\ell = 1. \tag{3.113}$$

Evaluating L.H.S. in the basis $\frac{(a_1^\dagger)^{n_1}(a_2^\dagger)^{n_2}}{\sqrt{n_1!n_2!}}|0\rangle, n_1 + n_2 = n$, we get after a choice of sign,

$$N_\ell = \sqrt{\frac{4\pi}{n+1}}\frac{(n-\ell)!(n+1)!\sqrt{(2\ell+1)!}}{n!\ell!(n+\ell+1)!} \tag{3.114}$$

and

$$T_\ell^\ell = \sqrt{\frac{4\pi}{n+1}}\frac{(n-\ell)!(n+1)!\sqrt{(2\ell+1)!}}{n!\ell!(n+\ell+1)!}a_2^{\dagger\ell}a_1^\ell. \tag{3.115}$$

Inserting (3.115) in (3.103) and using (3.111), we get, after a short calculation,

$$T_n(\ell)^{\frac{1}{2}} = \sqrt{\frac{4\pi}{n+1}\frac{n!(n+1)!}{(n-\ell)!(n+\ell+1)!}} \tag{3.116}$$

which gives $T_n(0)^{\frac{1}{2}} = \sqrt{\frac{4\pi}{n+1}}$ as claimed earlier.

Chapter 4

Scalar Fields on the Fuzzy Sphere

The free Euclidean action for the fuzzy sphere for a scalar field is

$$S_0 = \frac{1}{n+1} \text{Tr} \left[-\frac{1}{2}[L_i, \hat{\phi}][L_i, \hat{\phi}] + \frac{\mu^2}{2}\hat{\phi}^2 \right] \tag{4.1}$$

where we now hat all operators or $(n+1) \times (n+1)$ matrices.

As we saw in chapter 2, the scalar field can be expanded in terms of the polarization tensors $\hat{T}_m^\ell$,

$$\hat{\phi} = \sum_{\ell,m} \phi_{\ell m} \hat{T}_m^\ell , \tag{4.2}$$

where $\phi_{\ell m}$ are complex numbers. For concreteness, we will restrict our attention to Hermitian scalar fields $\hat{\phi}^\dagger = \hat{\phi}$. Since $(\hat{T}_m^\ell)^\dagger = (-1)^m \hat{T}_{-m}^\ell$, this implies that $\bar{\phi}_{\ell,m} = (-1)^m \phi_{\ell,-m}$.

In terms of $\phi_{\ell m}$'s, the action (4.1) is

$$S_0 = \sum_{\ell,m}^{n+1} \frac{|\phi_{\ell m}|^2}{2}(\ell(\ell+1) + \mu^2)$$

$$= \sum_{\ell=0}^{n+1} \frac{\phi_{\ell,0}^2}{2}(\ell(\ell+1) + \mu^2) + 2\sum_{\ell=0}^{n+1}\sum_{m=1}^{\ell} \frac{|\phi_{\ell m}|^2}{2}(\ell(\ell+1) + \mu^2). \tag{4.3}$$

The generating function for correlators in this model is

$$Z_0(\hat{J}) = \mathcal{N}_0 \int D\hat{\phi}\, e^{-S_0 + \frac{1}{n+1}\text{Tr}\hat{J}\hat{\phi}} \tag{4.4}$$

where $\hat{J}$, the "external current" is an $(n+1) \times (n+1)$ Hermitian matrix. Also

$$\mathcal{N}_0 = \left[\int D\hat{\phi}\, e^{-S_0} \right]^{-1} \tag{4.5}$$

is the usual normalization chosen so that

$$Z_0(0) = 1 \tag{4.6}$$

while

$$D\hat{\phi} = \prod_{\ell \leq n/2} \frac{d\phi_{\ell 0}}{\sqrt{2\pi}} \prod_{|m| \leq \ell} \frac{d\bar{\phi}_{\ell m} d\phi_{\ell m}}{2\pi i} . \tag{4.7}$$

Let us write

$$\hat{J} = \sum_{\ell,m} J_{\ell m} \hat{T}_m^\ell , \quad \bar{J}_{\ell m} = (-1)^m J_{\ell m} . \tag{4.8}$$

Then

$$\text{Tr}\hat{J}\hat{\phi} = \sum_{\ell,m} \bar{J}_{\ell m} \phi_{\ell m} = \sum_{\ell=0}^{n+1} J_{\ell 0}\phi_{\ell 0} + \sum_{\ell} \sum_{m \geq 1}^{\ell} (\bar{J}_{\ell m}\phi_{\ell m} + J_{\ell m}\bar{\phi}_{\ell m}) \tag{4.9}$$

and

$$Z_0(\hat{J}) = \mathcal{N}_0 \int d\hat{\phi} \exp\left[\sum_\ell \left(\frac{-\phi_{\ell,0}^2}{2}(\ell(\ell+1) + \mu^2) + J_{\ell 0}\phi_{\ell 0} \right) + \right.$$
$$\left. \sum_{\ell=0}^{n+1} \sum_{m=1}^{\ell} -|\phi_{\ell m}|^2(\ell(\ell+1)+\mu^2) + \bar{J}_{\ell m}\phi_{\ell m} + J_{\ell m}\bar{\phi}_{\ell m} \right] . \tag{4.10}$$

It is a product of Gaussians. Substituting

$$\phi_{\ell m} = \chi_{\ell m} + \frac{J_{\ell m}}{\ell(\ell+1) + \mu^2} \tag{4.11}$$

and fixing $\mathcal{N}_0$ by the condition $Z(0) = 1$, we get

$$Z_0(\hat{J}) = \prod_{\ell m} \exp\left[\frac{\bar{J}_{\ell m} J_{\ell m}}{2[\ell(\ell+1)+\mu^2]} \right] = \exp\left[\text{Tr}\frac{1}{2}\hat{J}^\dagger \frac{1}{(-\Delta+\mu^2)}\hat{J} \right] . \tag{4.12}$$

Using (4.10) and (4.12) we can compute all correlators (Schwinger functions) of ϕ's. For example,

$$\langle \bar{\phi}_{\ell' m'}\phi_{\ell m} \rangle := \mathcal{N}_0 \int D\hat{\phi}\bar{\phi}_{\ell' m'}\phi_{\ell m}e^{-S} = \left. \frac{\partial^2 Z_0(\hat{J})}{\partial J_{\ell' m'}\partial\bar{J}_{\ell m}} \right|_{J=0} = \frac{\delta_{\ell'\ell}\delta_{m'm}}{\ell(\ell+1)\mu^2} . \tag{4.13}$$

All the correlators of $\hat{\phi}$ follow from (4.13). For instance

$$\langle \hat{\phi}^2 \rangle = \sum_{\ell,m,\ell',m'} \hat{T}_{m'}^{\ell'\dagger}\hat{T}_m^\ell \langle \bar{\phi}_{\ell' m'}\phi_{\ell m} \rangle = \sum_{\ell,m} \frac{\hat{T}_m^\ell \hat{T}_m^{\dagger\ell}}{\ell(\ell+1)+\mu^2} . \tag{4.14}$$

From this follow the correlators under the coherent state or Weyl maps. The latter (or working with matrices) is more convenient for current purposes. We have not given $*_W$ explicitly earlier for S_F^2. But we will give the needed details here.

The image ϕ_W under the Weyl map of $\hat{\phi}$ has been defined earlier using the coherent state symbol ϕ_c of $\hat{\phi}$, $\phi_c(z)$ being $\langle z|\hat{\phi}|z\rangle$. Since $\hat{T}_m^\ell$ becomes Y_m^ℓ under the Weyl map, we get, using $\bar{Y}_m^\ell = (-1)^m Y_{-m}^\ell$, and dropping the subscript W,

$$\langle \phi(\vec{x})\phi(\vec{x}')\rangle \equiv G_n(\vec{x},\vec{x}') = \sum_{\ell=0}^{n}\sum_{m=-\ell}^{\ell} \frac{Y_m^\ell(\vec{x})\bar{Y}_m^\ell(\vec{x}')}{\ell(\ell+1)+\mu^2}$$

$$= \sum_{\ell=0}^{n}\sum_{m=-\ell}^{\ell} (-1)^m \frac{Y_m^\ell(\vec{x})Y_{-m}^\ell(\vec{x}')}{\ell(\ell+1)+\mu^2}. \qquad (4.15)$$

So as

$$(-1)^m = (-1)^{-m}, \qquad (4.16)$$

$$G_n(\vec{x},\vec{x}') = G_n(\vec{x}',\vec{x}). \qquad (4.17)$$

The symmetry of G_n is important for calculations.

4.1 Loop Expansion

There is a standard method to develop the loop expansion in the presence of interactions. Suppose the partition function is

$$Z(\hat{J}) = \mathcal{N} \int D\hat{\phi}\, e^{-S+\frac{1}{n+1}\mathrm{Tr}\hat{J}\hat{\phi}}, \qquad (4.18)$$

$$S = S_0 + \frac{1}{n+1}\frac{\lambda}{4!}\mathrm{Tr}\hat{\phi}^4 := S_0 + S_I, \quad \lambda > 0, \qquad (4.19)$$

$$\mathcal{N} = \left[\int D\hat{\phi}\, e^{-S}\right] \Rightarrow Z(0) = 1. \qquad (4.20)$$

Let

$$V(\ell_1 m_1; \ell_2 m_2; \ell_3 m_3; \ell_4 m_4) = \mathrm{Tr}\left(\hat{T}_{m_1}^{\ell_1}\hat{T}_{m_2}^{\ell_2}\hat{T}_{m_3}^{\ell_3}\hat{T}_{m_4}^{\ell_4}\right). \qquad (4.21)$$

We can further abbreviate L.H.S. as follows:

$$V(\ell_1 m_1; \ell_2 m_2; \ell_3 m_3; \ell_4 m_4) := V(1234). \qquad (4.22)$$

Now since

$$S_I = \frac{1}{n+1} \frac{\lambda}{4!} \mathrm{Tr}\left(\hat{T}^{\ell_1}_{m_1} \hat{T}^{\ell_2}_{m_2} \hat{T}^{\ell_3}_{m_3} \hat{T}^{\ell_4}_{m_4}\right) \phi_{\ell_1 m_1} \phi_{\ell_2 m_2} \phi_{\ell_3 m_3} \phi_{\ell_4 m_4}$$

$$\equiv \frac{\lambda}{4!} V(l_1, m_1; l_2, m_2; l_3, m_3; l_4, m_4; j) \phi_{\ell_1 m_1} \phi_{\ell_2 m_2} \phi_{\ell_3 m_3} \phi_{\ell_4 m_4}$$

$$\equiv \frac{\lambda}{4!} V(1234) \phi_{\ell_1 m_1} \phi_{\ell_2 m_2} \phi_{\ell_3 m_3} \phi_{\ell_4 m_4} , \qquad (4.23)$$

we can write, using (4.9),

$$Z(\hat{J}) = \mathcal{N} \exp\left[-\frac{\lambda}{4!} V(1234) \frac{\partial}{\partial \bar{J}_{\ell_1 m_1}} \frac{\partial}{\partial \bar{J}_{\ell_2 m_2}} \frac{\partial}{\partial \bar{J}_{\ell_3 m_3}} \frac{\partial}{\partial \bar{J}_{\ell_4 m_4}}\right] \times$$

$$\int D\hat{\phi} e^{-S_0 + \frac{1}{n+1} \mathrm{Tr} \hat{J} \hat{\phi}}$$

$$= \frac{\mathcal{N}}{\mathcal{N}_0} \exp\left[-\frac{\lambda}{4!} V(1234) \frac{\partial}{\partial \bar{J}_{\ell_1 m_1}} \frac{\partial}{\partial \bar{J}_{\ell_2 m_2}} \frac{\partial}{\partial \bar{J}_{\ell_3 m_3}} \frac{\partial}{\partial \bar{J}_{\ell_4 m_4}}\right]$$

$$\exp\left[\frac{1}{2} \sum_{\ell, m} \bar{J}_{\ell m} \frac{1}{-\Delta_\ell + \mu^2} J_{\ell m}\right],$$

$$(-\Delta_\ell + \mu^2)^{-1} = \frac{1}{\ell(\ell+1) + \mu^2}. \qquad (4.24)$$

Even before proceeding to calculate the one-loop two-point function, one can see that the interaction $V(1234)$ in (4.23) has invariance only under cyclic permutation of its factors ℓ_i, m_i and is not invariant under transpositions of adjacent factors. This means that we have to take care to distinguish between "planar" and "non-planar" graphs while doing perturbation theory as we shall see later below.

The function $V(1234)$ may be conveniently written as

$$V(1234) = (n+1) \prod_{i=1}^{4} (2\ell_i + 1)^{1/2} \times$$

$$\sum_{l,m}^{l=n} \begin{Bmatrix} \ell_1 & \ell_2 & l \\ \frac{n}{2} & \frac{n}{2} & \frac{n}{2} \end{Bmatrix} \begin{Bmatrix} \ell_3 & \ell_4 & l \\ \frac{n}{2} & \frac{n}{2} & \frac{n}{2} \end{Bmatrix} (-1)^m C^{\ell_1 \, \ell_2 \, l}_{m_1 m_2 m} C^{\ell_3 \, \ell_4 \, l}_{m_3 m_4 - m}. \qquad (4.25)$$

The $C^{l_1 \, l_2 \, l}_{m_1 m_2 m}$ are the Clebsch-Gordan (C-G) coefficients and the objects with 6 entries within brace brackets are the $6j$ symbols. Although less obvious, the R.H.S of (4.25) has cyclic symmetry, as can be verified using properties of $6j$ symbols and C-G coefficients.

The loop expansion of $Z(J)$ is its power series expansion in λ. By differentiating it with respect to the currents followed by setting them zero, we can generate the loop expansion of the correlators. The K-loop term is the λ^K-th term, the zero loop being referred to as the tree term. We can write

$$Z(J) = \sum_0^\infty \lambda^K z_K(J) \qquad (4.26)$$

where $\lambda^K z_K(J)$ is the K-loop term.

The factor $\mathcal{N}/\mathcal{N}_0$ contributes multiplicative vacuum fluctuation diagrams to the correlation functions. It is a common factor to *all* correlators, and is a phase in Minkowski (real time) regime.

4.2   The One-Loop Two-Point Function

Of particular interest is the one-loop two-point function where one can see a "non-planar" graph unique to noncommutative theories.

Expanding its numerator and denominator to $O(\lambda)$, we get for $Z(\hat{J})$,

$$Z(\hat{J}) \approx \left(1 - \frac{\lambda}{4!} V(1234) \frac{\partial}{\partial \bar{J}_{\ell_1 m_1}} \frac{\partial}{\partial \bar{J}_{\ell_2 m_2}} \frac{\partial}{\partial \bar{J}_{\ell_3 m_3}} \frac{\partial}{\partial \bar{J}_{\ell_4 m_4}} \right) \times$$

$$\exp\left[\frac{1}{2} \sum_{\ell,m} \bar{J}_{\ell m} \frac{1}{-\Delta_\ell + \mu^2} J_{\ell m} \right] \times$$

$$\left(1 - \frac{\lambda}{4!} V(1234) \langle \phi_{\ell_1 m_1} \phi_{\ell_2 m_2} \phi_{\ell_3 m_3} \phi_{\ell_4 m_4} \rangle \right)^{-1}. \qquad (4.27)$$

Here, the argument $i \in (1,2,3,4)$ in $V(1234)$ is to be interpreted as $\ell_i m_i$ and ℓ_i, m_i are to be summed over. Also the denominator comes from expanding $\mathcal{N}$ as power series in λ:

$$\mathcal{N} = \mathcal{N}(\lambda) := \sum_{K=0}^\infty \lambda^K \mathcal{N}_K. \qquad (4.28)$$

This contributes disconnected diagrams, two of which are planar and one is non-planar. The disconnected diagrams are precisely canceled by other terms of (4.24) as we shall see.

The $O(\lambda)$ term of (4.24) or (4.27) is $\lambda z_1(J)$ where

$$z_1(J) = \left[\frac{\mathcal{N}_1}{\mathcal{N}_0} - \frac{1}{4!} V(1234) \frac{\partial}{\partial \bar{J}_{\ell_1 m_1}} \frac{\partial}{\partial \bar{J}_{\ell_2 m_2}} \frac{\partial}{\partial \bar{J}_{\ell_3 m_3}} \frac{\partial}{\partial \bar{J}_{\ell_4 m_4}} \right] \times$$

$$\exp\left(\frac{1}{2} \sum_{\ell,m} \bar{J}_{\ell m} \frac{1}{-\Delta_\ell + \mu^2} J_{\ell m} \right), \qquad (4.29)$$

The two-point function follows by differentiation as in (4.13).

Expanding the exact two-point function $\langle \phi_{\ell m} \bar{\phi}_{\ell' m'} \rangle$ in powers of λ,

$$\langle \phi_{\ell m} \bar{\phi}_{\ell' m'} \rangle = \langle \phi_{\ell m} \bar{\phi}_{\ell' m'} \rangle_0 + \lambda \langle \phi_{\ell m} \bar{\phi}_{\ell' m'} \rangle_1 + \dots \tag{4.30}$$

we get

$$\langle \phi_{\ell m} \bar{\phi}_{\ell' m'} \rangle_1 = \frac{\partial}{\partial \bar{J}_{\ell m} \partial J_{\ell' m'}} z_1(J) \Big|_{J=0}$$

$$= \frac{\mathcal{N}_1}{\mathcal{N}_0} \langle \phi_{\ell m} \bar{\phi}_{\ell' m'} \rangle_0 - \left(\frac{\partial}{\partial \bar{J}_{\ell m}} \frac{\partial}{\partial J_{\ell' m'}} \frac{\partial}{\partial J_{\ell_1 m_1}} \frac{\partial}{\partial J_{\ell_2 m_2}} \frac{\partial}{\partial \bar{J}_{\ell_3 m_3}} \frac{\partial}{\partial \bar{J}_{\ell_4 m_4}} \right) \times$$

$$\left[\frac{\lambda}{4!} V(1234) \exp \left(\frac{1}{2} \sum_{\ell,m} \bar{J}_{\ell m} \frac{1}{-\Delta_\ell + \mu^2} J_{\ell m} \right) \right]_{J=0} \tag{4.31}$$

(4.31) has both disconnected and connected diagrams. We briefly examine them.

i. Disconnected Diagrams:

They come when the differentiations $\frac{\partial}{\partial J_{\ell' m'}}$, $\frac{\partial}{\partial \bar{J}_{\ell m}}$ both hit the same factor in the product of (4.31) to produce the free propagator. There are three such terms, two of which are planar diagrams and a non-planar diagram. These add up to $-\left(\frac{\mathcal{N}_1}{\mathcal{N}_0} \right) [-\Delta_\ell + \mu^2]^{-1} \delta_{\ell \ell'} \delta_{mm'}$:

$$\langle \phi_{\ell m} \bar{\phi}_{\ell' m'} \rangle_1^D = -\frac{\mathcal{N}_1}{\mathcal{N}_0} [\Delta_\ell + \mu^2]^{-1} \delta_{\ell \ell'} \delta_{mm'} \tag{4.32}$$

thus canceling the first term of (4.31).

ii. Connected Diagrams:

They arise when the differentiation on external currents is applied to different factors in the product. There are $4 \times 3 = 12$ such terms, giving

$$\langle \phi_{\ell m} \bar{\phi}_{\ell' m'} \rangle_1^C = -\frac{\lambda}{4!} \times$$

$$\left[8 \frac{\delta_{\ell \ell_4} \delta_{m+m_4,0} (-1)^{m_4}}{-\Delta_\ell + \mu^2} \frac{\delta_{\ell' \ell_3} \delta_{m+m_3,0} (-1)^{m_3}}{-\Delta_{\ell'} + \mu^2} \frac{\delta_{\ell_1 \ell_2} \delta_{m_1+m_2,0} (-1)^{m_2}}{-\Delta_{\ell_1} + \mu^2} V(1234) + \right.$$

$$\left. 4 \frac{\delta_{\ell \ell_2} \delta_{m+m_2,0} (-1)^{m_2}}{-\Delta_\ell + \mu^2} \frac{\delta_{\ell' \ell_4} \delta_{m+m_4,0} (-1)^{m_4}}{-\Delta_{\ell'} + \mu^2} \frac{\delta_{\ell_1 \ell_3} \delta_{m_1+m_3,0} (-1)^{m_3}}{-\Delta_{\ell_1} + \mu^2} V(1234) \right] \tag{4.33}$$

where, keeping in mind the symmetries of the trace, we have decomposed (4.33) into planar and nonplanar contributions. In the planar case, the indices of adjacent $\hat{T}$'s get contracted. There are 8 such terms. In the non-planar case, it is the indices of the alternate $\hat{T}$'s that get contracted, and there are 4 such terms.

The planar term can be further simplified, by observing that $\hat{T}^{\ell_1}_{m_1}\hat{T}^{\ell_1}_{-m_1}(-1)^{m_1} = \hat{T}^{\ell_1}_{m_1}\hat{T}^{\ell_1\dagger}_{m_1}$ is rotationally invariant, and thus proportional to $\mathbf{1}$, the constant of proportionality being $1/(n+1)$ (as seen by taking the trace). Incising the external legs, the one-loop planar contribution is thus

$$(-\Delta_\ell + \mu^2)^{-1}\langle\phi_{\ell m}\bar{\phi}_{\ell' m'}\rangle_1^{C,planar}(-\Delta_{\ell'} + \mu^2)^{-1} =$$

$$-\frac{1}{3}\delta_{\ell\ell'}\delta_{m+m',0}(-1)^m\sum_{\ell=0}^{n}\frac{2\ell+1}{\ell(\ell+1)+\mu^2}. \quad (4.34)$$

In the non-planar case, the indices of nonadjacent $\hat{T}$'s get contracted. To evaluate the non-planar term, we need to make explicit use of the form (4.25). There are four such terms giving

$$(-\Delta_\ell + \mu^2)^{-1}\langle\phi_{\ell m}\bar{\phi}_{\ell' m'}\rangle_1^{C,nonplanar}(-\Delta_{\ell'} + \mu^2)^{-1} =$$

$$-\frac{1}{6}(n+1)\sum_{\ell_1,m_1,\ell_3,m_3}\prod_{i=1}^{4}(2\ell_i+1)^{1/2}\sum_{l,m}\begin{Bmatrix}\ell_1 & \ell_2 & l \\ \frac{n}{2} & \frac{n}{2} & \frac{n}{2}\end{Bmatrix}\begin{Bmatrix}\ell_3 & \ell_4 & l \\ \frac{n}{2} & \frac{n}{2} & \frac{n}{2}\end{Bmatrix}\times$$

$$\times(-1)^m C^{\ell_1\ \ell_2\ l}_{m_1 m_2 m}C^{\ell_3\ \ell_4\ l}_{m_3 m_4 -m}\frac{\delta_{\ell_1\ell_3}\delta_{m_1+m_3,0}(-1)^{m_3}}{\ell_1(\ell_1+1)+\mu^2},$$

$$= -\frac{1}{6}(n+1)\sqrt{(2\ell_2+1)(2\ell_4+1)}\sum_{\ell,m,\ell_1,m_1}(2\ell+1)\begin{Bmatrix}\ell_1 & \ell_2 & l \\ \frac{n}{2} & \frac{n}{2} & \frac{n}{2}\end{Bmatrix}\begin{Bmatrix}\ell_1 & \ell_4 & l \\ \frac{n}{2} & \frac{n}{2} & \frac{n}{2}\end{Bmatrix}\times$$

$$\times(-1)^{m-m_1}C^{\ell_1\ \ell_2\ l}_{m_1 m_2 m}C^{\ell_1\ \ell_4\ l}_{-m_1 m_4 -m} \quad (4.35)$$

We first perform the sum

$$\sum_{m,m_1}(-1)^{m-m_1}C^{\ell_1\ \ell_2\ l}_{m_1 m_2 m}C^{\ell_1\ \ell_4\ l}_{-m_1 m_4 -m} \quad (4.36)$$

for which we need the identities

$$C^{\ell_1\ \ell_2\ l}_{m_1 m_2 m} = (-1)^{\ell_1-m_1}\sqrt{\frac{2\ell+1}{2\ell_2+1}}C^{\ell_1\ l\ \ell_2}_{m_1 -m -m_2}, \quad (4.37)$$

$$C^{\ell_1\ \ell_4\ l}_{-m_1 m_4 -m} = (-1)^{\ell-\ell_4+m_1}\sqrt{\frac{2\ell+1}{2\ell_2+1}}C^{\ell_1\ l\ \ell_4}_{m_1 -m m_4}, \quad (4.38)$$

$$\sum_{m_1,m_2}C^{\ell_1\ \ell_2\ \ell_3}_{m_1 m_2 m_3}C^{\ell_1\ \ell_2\ \ell_4}_{m_1 m_2 m_4} = \delta_{\ell_3\ell_4}\delta_{m_3 m_4}. \quad (4.39)$$

This simplifies the non-planar contribution to

$$(-\Delta_\ell+\mu^2)^{-1}\langle\phi_{\ell m}\bar{\phi}_{\ell' m'}\rangle_1^{C,nonplanar}(-\Delta_{\ell'}+\mu^2)^{-1} = -\frac{1}{6}(n+1)\delta_{\ell_2\ell_4}\delta_{m_2+m_4}\times$$

$$(-1)^{m_2-\ell_2}\sum_{\ell,\ell_1}(-1)^{\ell_1+\ell}\frac{(2\ell+1)(2\ell_1+1)}{\ell_1(\ell_1+1)+\mu^2}\begin{Bmatrix}\ell_1 & \ell_2 & l \\ \frac{n}{2} & \frac{n}{2} & \frac{n}{2}\end{Bmatrix}\begin{Bmatrix}\ell_1 & \ell_4 & l \\ \frac{n}{2} & \frac{n}{2} & \frac{n}{2}\end{Bmatrix}. \quad (4.40)$$

This can be simplified even further, using the following identity involving the $6j$ symbols:

$$\sum_{\ell}(-1)^{n+\ell}(2\ell+1)\begin{Bmatrix}\ell_1 & \ell_2 & l \\ \frac{n}{2} & \frac{n}{2} & \frac{n}{2}\end{Bmatrix}\begin{Bmatrix}\ell_1 & \ell_4 & l \\ \frac{n}{2} & \frac{n}{2} & \frac{n}{2}\end{Bmatrix}=\begin{Bmatrix}\ell_1 & \frac{n}{2} & \frac{n}{2} \\ \ell_4 & \frac{n}{2} & \frac{n}{2}\end{Bmatrix}. \tag{4.41}$$

We finally get

$$(-\Delta_\ell+\mu^2)^{-1}\langle\phi_{\ell m}\bar\phi_{\ell'm'}\rangle_1^{C,nonplanar}(-\Delta_{\ell'}+\mu^2)^{-1}=$$
$$-\frac{1}{6}(n+1)\delta_{\ell_2\ell_4}\delta_{m_2+m_4}(-1)^{m_2}(-1)^{\ell_4+n}\times$$
$$\sum_{\ell_1}(-1)^{\ell_1}\frac{(n+1)(2\ell_1+1)}{\ell_1(\ell_1+1)+\mu^2}\begin{Bmatrix}\ell_1 & \frac{n}{2} & \frac{n}{2} \\ \ell_4 & \frac{n}{2} & \frac{n}{2}\end{Bmatrix} \tag{4.42}$$

The surprising fact is that this nonplanar contribution to the one-loop two-point function does not vanish even in the limit of $n\to\infty$ [16]. In particular the difference between planar and non-planar contributions remains finite. To see this, we can use the Racah formula [46]

$$\begin{Bmatrix}\ell_1 & \frac{n}{2} & \frac{n}{2} \\ \ell_4 & \frac{n}{2} & \frac{n}{2}\end{Bmatrix}\simeq\frac{(-1)^{\ell_1+\ell_4+n}}{n}P_{\ell_1}\left(1-\frac{2\ell_4^2}{n^2}\right) \tag{4.43}$$

where P_ℓ are the usual Legendre polynomials. Recall that the planar contribution from each Feynman diagram is proportional to

$$\sum_{\ell=0}\frac{2\ell+1}{\ell(\ell+1)+\mu^2} \tag{4.44}$$

which is logarithmically divergent. The *difference*, proportional to

$$\delta\equiv\sum_{\ell_1=0}\frac{2\ell_1+1}{\ell_1(\ell_1+1)+\mu^2}-\sum_{\ell_1}(-1)^{\ell_1}\frac{(n+1)(2\ell_1+1)}{\ell_1(\ell_1+1)+\mu^2}\begin{Bmatrix}\ell_1 & \frac{n}{2} & \frac{n}{2} \\ \ell_4 & \frac{n}{2} & \frac{n}{2}\end{Bmatrix} \tag{4.45}$$

between planar and nonplanar terms then simplifies to

$$\delta=\sum_{\ell=0}^{n}\frac{2\ell+1}{\ell(\ell+1)+\mu^2}\left[1-P_{\ell_1}\left(1-\frac{2\ell^2}{n^2}\right)\right]. \tag{4.46}$$

Using $y=\ell/n$, this sum is easily approximated by an integral for large n. Finally, substituting $1-y^2/2=x$ gives us

$$\delta\simeq\int_{-1}^{1}dx\frac{1-P_{\ell_1}(x)}{1-x}=2\sum_{k=1}^{\ell_1}\left(\frac{1}{k}\right) \tag{4.47}$$

This incorporates the UV-IR mixing [16–18]: δ depends on the external momentum ℓ_1 is non-vanishing for even small values of ℓ_1. Thus integrating out high energy (or UV) modes in the loop produces non-trivial effects even at low (or IR) external momenta.

This mixing has the potential to pose a serious challenge to any lattice program that uses matrix models on S_F^2 to discretize continuum models on the sphere. It is therefore important to ask if its effect can effectively be restricted to a class of n-point functions. To this end, one can calculate the four-point function at one-loop. Interestingly in this case, careful analysis shows that the difference between planar and the non-planar diagrams vanishes in the limit of large n [18]. Since only the quadratic term is affected by UV-IR mixing (albeit by a complicated momentum dependence), it suggests that appropriately "normal-ordered" vertices may completely eliminate this problem. That this is indeed the case was shown by Dolan, O'Connor and Presnajder [18]. Working with a modified action

$$S_0 = \frac{1}{n+1}\text{Tr}\left[-\frac{1}{2}[L_i, \hat{\phi}][L_i, \hat{\phi}] + \frac{\mu^2}{2}\hat{\phi}^2 + \frac{\lambda}{4!} : \hat{\phi}^4 :\right] \qquad (4.48)$$

where

$$\text{Tr} : \hat{\phi}^4 := \text{Tr}\left[\hat{\phi}^4 - 12\sum_{\ell,m}\frac{\hat{\phi}\hat{T}_{\ell m}^\dagger\hat{T}_{\ell m}\hat{\phi}}{\ell(\ell+1)+\mu^2} + 2\sum_{\ell,m}\frac{[\hat{\phi},\hat{T}_{\ell m}]^\dagger[\hat{\phi},\hat{T}_{\ell m}]}{\ell(\ell+1)+\mu^2}\right],$$
$$(4.49)$$

they showed that one gets the standard action on the sphere in the continuum limit $n \to \infty$.

One may ask if normal-ordering can help cure the UV-IR mixing problem in higher dimensions, say, on $S_F^2 \times S_F^2$. Here the problem is much more severe, and unfortunately persists after normal-ordering [19, 20].

4.3 Numerical Simulations

One of the principal motivations of studying the fuzzy sphere is to approximate the ordinary sphere without losing any of the symmetries. We will briefly indicate the issues involved here, drawing from [21].

Functions are approximated by $(n+1) \times (n+1)$ matrices, and we are interested in a quantum partition function constructed from these matrices. For example, while studying the action (4.19), one may be interested in expectation values of some operator $h(\hat{\phi})$:

$$\langle h(\hat{\phi})\rangle = \frac{1}{Z}\int D\hat{\phi}e^{-S[\hat{\phi}]}h(\hat{\phi}). \qquad (4.50)$$

For Hermitian matrices $\hat{\phi}$, the above integral involves $(n+1)^2$ integrations. Were one to use a typical integration routine which requires p steps, calculation of the partition function would require $p^{(n+1)^2}$ steps. Since this diverges exponentially with matrix size, we require another strategy.

Notice that in the exponential in the definition of the partition function, only a small number of configurations of $\hat{\phi}$ makes a significant contribution to the integral. If there were a way to focus on these regions of the configuration space, the computation could be speeded up considerably. To this end, if the matrix entries of $\hat{\phi}$ could be generated randomly with the probability distribution $p(\hat{\phi}) = e^{-S[\hat{\phi}]}/Z$, then the expectation value of $h(\hat{\phi})$ can be calculated simply is a simple average of a random number of "draws" of the field $\hat{\phi}$. The nice thing about this strategy is that since $\hat{\phi}$ is drawn according to the probability distribution, it would typically be drawn from regions where the probability is large, and hence make a significant contribution to the integral.

The fields (or matrices) $\hat{\phi}$ are randomly chosen (with distribution $p(\hat{\phi})$) using a one-step Markov process. Starting from an initial configuration, a sequence of matrices is drawn inductively, using the Monte-Carlo Metropolis method. For example, at the $(k+1)$-th step, a matrix ψ is drawn randomly from a uniform distribution near the random matrix ψ_k obtained at step k. One then calculates difference between the actions $\delta S \equiv S[\psi] - S[\psi_k]$. The matrix ψ is accepted as the next one in the sequence with probability $min(e^{-\delta S}, 1)$, otherwise one retains ψ_k as ψ_{k+1}.

After a sufficiently large number of such steps, one is reasonably sure that the matrices generated in this way are sufficiently random. Having thus created a large bank of such random matrices, one can then explicitly calculate the statistical average of any operator of interest as

$$\langle h(\hat{\phi}) \rangle = \frac{1}{N} \sum_N h(\hat{\phi}_i) \qquad (4.51)$$

By judiciously choosing the operator $h(\hat{\phi})$, we can investigate various physical phenomena. For example, the *finite volume susceptibility* given by

$$\chi \equiv \langle \text{Tr } \hat{\phi}^2 \rangle - \langle |\text{Tr } \hat{\phi}| \rangle^2, \qquad (4.52)$$

which diverges in the continuum theory at a continuous phase transition, is one such indicator: sharp peaks in susceptibility will separate regions of different phases.

We will briefly discuss the results for 3 situations:

1. Scalar field theory with quartic interaction on fuzzy S^2.

2. Scalar field theory with quartic interaction on fuzzy S^2 and discrete Euclidean time.

3. Gauge theory on fuzzy $S^2 \times S^2$.

4.3.1 *Scalar Field Theory on Fuzzy S^2*

Following [22], the action of interest is

$$S[\Phi] = \mathrm{Tr}\left[a\Phi^\dagger[L_i,[L_i,\Phi]] + b\Phi^2 + c\Phi^4\right]. \tag{4.53}$$

Three phases of this model have been identified:

1. **Disordered:** For small negative values of b, the fluctuations fluctuate in the neighborhood of $\Phi = 0$: $\langle\mathrm{Tr}\ \Phi\rangle = 0 = \langle\mathrm{Tr}\ \Phi^2\rangle$.

2. **Non-uniform ordered:** As b becomes more negative, there develop symmetrical peaks for the probability distribution of $\langle\mathrm{Tr}\ \Phi\rangle$. In particular, there is evidence of a spontaneous breakdown of rotational invariance.

3. **Uniform ordered:** For large negative values of b, the probability distribution of $\langle\mathrm{Tr}\ \Phi\rangle$ are located approximately at $\pm(n+1)\sqrt{-b/c}$. In addition, $\langle\mathrm{Tr}\ \Phi^2\rangle \simeq \langle\mathrm{Tr}\ \Phi\rangle/(n+1)$, indicating that $\Phi \simeq \sqrt{-b/c}\mathbf{1}$, so that rotational symmetry is again restored.

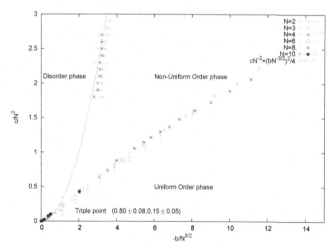

Fig. 4.1 Phase diagram obtained from Monte–Carlo simulations of the model (4.53), from Garcia Flores et. al. [22]

In the limit of $a \to 0$ or equivalently, if $|b|, c \gg |a|$, the model reduces to the usual random matrix model, which has a third order phase transition between the disordered and non-uniform ordered phase at $c = b^2/4N$ [24]. Again, the numerical evidence confirms this behavior.

When (discretised) Euclidean time is included, numerical results again confirm the existence of the abovementioned three phases [23]. One important advantage of this model is that the radius R of the fuzzy sphere is now an independent parameter, and novel and interesting scaling limits are now accessible. For example, keeping R fixed, $n \to \infty$ gives the fuzzy sphere, taking both R and n to infinity but with $R^2/n \to 0$ gives the commutative plane, while keeping R^2/n finite and non-zero when R and $n \to \infty$ gives the noncommutative plane.

4.3.2　*Gauge Theory on Fuzzy $S^2 \times S^2$*

Another class of models closely related to the fuzzy sphere comes from the action

$$S[A] = n\mathrm{Tr}\left(\frac{1}{4}[A_\mu, A_\nu]^2 + \frac{2i}{3}\alpha f_{\mu\nu\rho}A_\mu A_\nu A_\rho\right) \qquad (4.54)$$

where A_μ are $n \times n$ traceless Hermitian matrices, and $f_{\mu\nu\rho}$ is a totally antisymmetric tensor. The equation of motion is

$$[A_\nu, [A_\mu, A_\nu]] - i\alpha f_{\mu\nu\rho}[A_\nu, A_\rho] = 0. \qquad (4.55)$$

Theoretical interest in these models in two-fold. Firstly, they arise as limits of certain nonperturbative formulations of string theory (see for eg [141] and references therein). Secondly, these models provide a potential regularization of Yang-Mills theories, and are thus amenable to numerical study.

For $f_{\mu\nu\rho} = \epsilon_{\mu\nu\rho}$, it is easy to see that the fuzzy sphere $A_\mu = \alpha L_\mu$ is a solution, apart from the trivial solution $A_\mu = 0$. Similarly, $f_{\mu\nu\rho} = \epsilon_{\mu\nu\rho}$ for $\mu, \nu, \rho \in \{1, 2, 3\}$ or $\{4, 5, 6\}$ and zero otherwise, one obtains the fuzzy $S^2 \times S^2$ as a solution.

These models hold out the possibility of studying topology changes as phase transitions [25–28]. Their results indicate that for α smaller than a critical value, the true ground state of the theory is A_μ equal to diagonal matrices. For α above the critical value, it is the fuzzy S^2 that is the ground state, while fuzzy $S^2 \times S^2$ can exist as a metastable state.

We refer to the original papers for further details.

Chapter 5

Instantons, Monopoles and Projective Modules

The two-sphere S^2 admits many nontrivial field configurations.

One such configuration is the instanton. It occurs when S^2 is Euclidean space-time. It is of particular importance as a configuration which tunnels between distinct "classical vacua" of a $U(1)$ gauge theory. An instanton can be regarded as the curvature of a connection for a $U(1)$-bundle on S^2. As there are an infinite number of $U(1)$-bundles on S^2 characterized by an integer k (Chern number), there are accordingly an infinite number of instantons as well.

We can also think of S^2 as the spatial slice of space-time $S^2 \times \mathbb{R}$. In that case, the instantons become monopoles. (The monopoles can be visualized as sitting at the center of the sphere embedded in $\mathbb{R}^3$. If a charged particle moves in its field, k is the product of its electric charge and monopole charge [45, 66].).

In algebraic language, what substitutes for bundles are "projective modules" [3]. Here we describe what they mean and find them for monopoles and instantons.

5.1 Free Modules, Projective Modules

Consider $Mat(N + 1) = Mat(2L + 1)$. It carries the left- and right-regular representations of the fuzzy algebra. Thus for each $a \in Mat(2L + 1)$ there are two operators a^L and a^R acting on $Mat(2L + 1)$ (thought of as a vector space) defined by

$$a^L b = ab, \quad a^R b = ba, \quad b \in Mat(N + 1) \tag{5.1}$$

with $a^L b^L = (ab)^L$ and $a^R b^R = (ba)^R$.

Definition: A module V for an algebra $\mathcal{A}$ is a vector space which carries a representation of $\mathcal{A}$.

Thus $V = Mat(N + 1)$ is an $\mathcal{A}$- ($= Mat(N + 1)-$) module. As this V carries two actions of $\mathcal{A}$, it is a bimodule. (But note that $a^R b^R = (ba)^R$.)

For an $\mathcal{A}$-module, linear combinations of vectors in V can be taken with coefficients in $\mathcal{A}$. Thus if $v_i \in V$ and $a_i \in \mathcal{A}$, $a_i v_i \in V$. A vector space over complex numbers in this language is a $\mathbb{C}$-module.

We consider only $\mathcal{A}$-modules V whose elements are finite-dimensional vectors $v_i = (v_{i1}, \cdots v_{iK})$ with $v_{ij} \in \mathcal{A}$. The action of $a \in \mathcal{A}$ on V is then $v_i \rightarrow av_i = (av_{i1}, \cdots, av_{ik})$.

Consider the identity $\mathbf{1}$ belonging to $V = Mat(N + 1)$. Then all its elements can be got by (left- or right-) $\mathcal{A}$-action. As an $\mathcal{A}$-module, it is one-dimensional. It is also "generated" by $\mathbf{1}$ as an $\mathcal{A}$-module. It is a "free" module as it has a basis.

Generally, an $\mathcal{A}$-module V is said to be free if it has a basis $\{e_i\}$, $e_i \in V$. That means that any $x \in V$ can be uniquely written as $\sum a_i e_i, a_i \in \mathcal{A}$. Uniqueness implies linear independence: $\sum a_i e_i = 0 \Leftrightarrow$ all $a_i = 0$.

The phrase "free" merits comment. It just means that there is no (additional) condition of the form $b_i e_i = 0, b_i \in \mathcal{A}$, with at least one $b_j \neq 0$. In other words, $\{e_i\}$ is a basis.

A class of free $Mat(N + 1)$-bimodules we can construct from $V = Mat(N+1)$ are $V \otimes \mathbb{C}^K \equiv V^K$. Elements of V^K are $v := (v_1, \ldots v_K), v_i \in V$. The left- and right- actions of $a \in \mathcal{A}$ on V^K are the natural ones: $a^L v = (av_1, \ldots, av_K), a^R v = (v_1 a, \ldots, v_K a)$.

V^K is a free module as it has the basis $\langle \{e_i\} : e_i = (0, \ldots, 0, \underbrace{1}_{i^{th} entry}, 0, \ldots, 0) \rangle$.

A projector P on the $\mathcal{A}$-module V^K is a $K \times K$ matrix $P = (P_{ij})$ with entries $P_{ij} \in \mathcal{A}$, fulfilling $P^\dagger = P, P^2 = P$ where $P_{ij}^\dagger = P_{ji}^*$. Consider PV^K. (We can also apply P on the right: $\xi \in V^K P \Rightarrow \xi_i = \xi_j P_{ji}$). On PV^K, we can generally act only on the right with $\mathcal{A}$, so it is only a right-$\mathcal{A}$-module and not a left one.

Any vector in PV^K is a linear combination of Pe_i with coefficients in $\mathcal{A}$ (acting on the right): $\xi \in PV^K \Rightarrow \xi = \sum_i (Pe_i)a_i, a \in \mathcal{A}$. But $\{Pe_i = f_i\}$ cannot be regarded as a basis as f_i are not linearly independent. There exist $a_i \in \mathcal{A}$, not all equal to zero, such that $\sum_i Pe_i a_i = 0$, that is, $\sum e_i a_i$ is in the kernel of P, without $\sum e_i a_i$ being 0. PV^K is an example of a projective module.

A module, projective or otherwise, is said to be trivial if it is a free module.

Note that PV^K is a summand in the decomposition $V^K = PV^K \oplus (1 - P)V^K$ of the trivial module V^K.

These ideas are valid (with possible technical qualifications) for any algebra $\mathcal{A}$ and an $\mathcal{A}$-module V. In particular, they are valid if $\mathcal{A}$ is the commutative algebra $C^\infty(M)$ of smooth functions on a manifold with point-wise multiplication. We now show that the elements of $\mathcal{A}$-modules are sections of bundles on M, picking $M = S^2$ for concreteness. In this picture, sections of twisted bundles on S^2, such as twisted $U(1)$-bundles, are elements of nontrivial projective modules. Such sections have a natural interpretation as charge-monopole wave functions.

It is a theorem of Serre and Swan [36] that all such sections can be obtained from projective modules using preceding algebraic constructions.

5.2   Projective Modules on $\mathcal{A} = C^\infty(S^2)$

Consider the free module $\mathcal{A}^2 = \mathcal{A} \otimes \mathbb{C}^2$. If $\hat{x}$ is the coordinate function, $(\hat{x}_i a)(x) = x_i a(x), a \in \mathcal{A}$, we can define the projector

$$P^{(1)} = \frac{1 + \vec{\tau} \cdot \hat{x}}{2} \tag{5.2}$$

where τ_i are the Pauli matrices. $P^{(1)}\mathcal{A}^2$ is an example of a projective module. $P^{(1)}\mathcal{A}^2$ carries an $\mathcal{A}$-action, left- and right- actions being the same.

The projector $P^{(1)}$ occurs routinely when discussing the charge-monopole system [70, 71] or the Berry phase [67]. We will now establish that $P^{(1)}\mathcal{A}^2$ is a nontrivial projective module. Its elements are known to be the wave functions for Chern number k (= product of electric and magnetic charges) $= 1$. For $k = -1$, we can use the projector $P^{(-1)} = \frac{1 - \vec{\tau} \cdot \hat{x}}{2}$.

At each x, $P^{(1)}(x)$ is of rank 1. If $P^{(1)}\mathcal{A}^2$ has a basis e, then $e(x)$ is an eigenstate of $P^{(1)}(x)$, $P^{(1)}(x)e(x) = e(x)$, and smooth in x. But there is no such e. For suppose that is not so. Let us normalize $e(x)$: $e^\dagger(x)e(x) = 1$. Let $f_a = \epsilon_{ab}e_b(\varepsilon_{ab} = -\varepsilon_{ba}, \varepsilon_{12} = +1)$. Then f is a smooth normalized vector perpendicular to e and annihilated by $P^{(1)}$: $P^{(1)}f = 0$. The operator

$$U = \begin{pmatrix} e_1 & f_1 \\ e_2 & f_2 \end{pmatrix}. \tag{5.3}$$

is unitary at each x $(U^\dagger(x)U(x) = 1)$ and

$$U^\dagger P^{(1)} U = \frac{1 + \tau_3}{2} \tag{5.4}$$

So we have rotated the hedgehog (winding number 1) map $\hat{x} : x \rightarrow \hat{x}(x)$ to the constant map $x \rightarrow (0,0,1)$. As that is impossible [45], e does not exist.

For higher k, we can proceed as follows. Take k copies of $\mathbb{C}^2$ and consider $\mathbb{C}^{2^k} = \mathbb{C}^2 \otimes \cdots \otimes \mathbb{C}^2$. Let $\vec{\tau}^{(i)}$ be the Pauli matrices acting on the ith slot in $\mathbb{C}^{2^k}$. That is $\vec{\tau}^{(i)} = 1 \otimes \cdots \otimes \vec{\tau} \otimes \cdots \otimes 1$. Then the projector for k is

$$P^{(k)} = \prod_{i=1}^{k} \frac{1 + \vec{\tau}^{(i)} \cdot \hat{x}}{2} \qquad (5.5)$$

and the projective module is

$$P^{(k)}[\mathcal{A} \otimes \mathbb{C}^{2^k}] := P^{(k)} \mathcal{A}^{2^k}. \qquad (5.6)$$

For $k = -|k|$, the projector in (5.5) gets replaced by

$$P^{(-|k|)} = \prod_{i=1}^{|k|} \frac{1 - \vec{\tau}^{(i)} \cdot \hat{x}}{2}. \qquad (5.7)$$

We can also construct the modules in another way. Let $k > 0$. Consider $z = (z_1, z_2)$ with $\sum_i |z_i|^2 = 1$. These are the z's of Chapter 2. For $k > 0$, let

$$v_k(z) = \frac{1}{\sqrt{Z_k}} \begin{pmatrix} z_1^k \\ z_2^k \end{pmatrix}, \quad Z_k = \sum_i |z_i|^{2k}. \qquad (5.8)$$

It is legitimate to put Z_k in the denominator: it cannot vanish without both z_i being 0, and that is not possible. $v_k(z)$ is normalized:

$$v_k^\dagger(z) v_k(z) = 1. \qquad (5.9)$$

So $v_k(z) \otimes v_k^\dagger(z)$ is a projector. Under $z_i \rightarrow z_i e^{i\theta}$, $v_k(z) \rightarrow v_k(z) e^{ik\theta}$ and the projector is invariant, so it depends only on $x = z^\dagger \vec{\tau} z \in S^2$. In this way, we get the projector $P'^{(k)}$

$$P'^{(k)}(x) = v_k(z) \otimes v_k^\dagger(\bar{z}) \qquad (5.10)$$

For $k = -|k| < 0$, such a projector is

$$P'^{(-|k|)}(x) = \bar{v}_{|k|}(\bar{z}) \otimes \bar{v}_{|k|}^\dagger(z) \qquad (5.11)$$

The projectors (5.10 , 5.11) are sometimes refered to as "Bott" projectors.

5.3 Equivalence of Projective Modules

We briefly explain the sense in which the projectors $P^{(k)}, P'^{(k)}$ and the modules $P^{(k)} \mathcal{A}^{2^k}$ and $P'^{(k)} \mathcal{A}^2$ are equivalent.

Two modules are said to be equivalent if the corresponding projectors are equivalent. But there are several definitions of equivalence of projectors [68]. We pick one which appears best for physics.

The $2^{2^k} \times 2^{2^k}$ matrix $P^{(k)}$ or the 2×2 matrix $P'^{(k)}$ can be embedded in the space of linear operators on an infinite-dimensional Hilbert space $\mathcal{H}$. The elements of $\mathcal{H}$ consist of $a = (a_1, a_2, \ldots), a_i \in C^\infty(S^2)$. The scalar product for $\mathcal{H}$ is $(b, a) = \int_{S^2} d\Omega \sum_l b_l^*(x) a_l(x)$. $\mathcal{H}$ is clearly an $\mathcal{A}$-module.

The embedding is accomplished by putting $P^{(k)}$ and $P'^{(k)}$ in the top left- corner of an "$\infty \times \infty$" matrix. The result is

$$\mathcal{P}^{(k)} = \begin{pmatrix} P^{(k)} & 0 \\ 0 & 0 \end{pmatrix}, \quad \mathcal{P}'^{(k)} = \begin{pmatrix} P'^{(k)} & 0 \\ 0 & 0 \end{pmatrix}. \tag{5.12}$$

A matrix U acting on $\mathcal{H}$ has "coefficients" in $\mathcal{A} : U_{ij} \in C^\infty(S^2)$. It is said to be unitary if $U^\dagger U = \mathbf{1}$ where each diagonal entry in $\mathbf{1}$ is the constant function on S^2 with value $1 \in \mathbb{C}$.

The projectors $P^{(k)}$ and $P'^{(k)}$ are said to be equivalent if there exists a unitary U such that $U \mathcal{P}^{(k)} U^\dagger = \mathcal{P}'^{(k)}$. If there is such a U, then $U \mathcal{P}^{(k)} a = \mathcal{P}'^{(k)} U a, a \in \mathcal{H}$. That means that wave functions given by $\mathcal{P}^{(k)} \mathcal{H}$ and $\mathcal{P}'^{(k)} \mathcal{H}$ are unitarily related. It is then reasonable to regard $P^{(k)} \mathcal{A}^{2^k}$ and $P'^{(k)} \mathcal{A}^2$ as equivalent.

Illustration:

We now illustrate this notion of equivalence using $P^{(k)}$ and $P'^{(k)}$. Since $P^{(\pm 1)} = P'^{(\pm 1)}$, $k = \pm 2$ is the first nontrivial example.

Let z_i be as above. Then the matrix with components $z_i \bar{z}_j$ is a projector. It is invariant under $z_i \to z_i e^{i\theta}$ and is a function of x. In fact

$$P^{(1)}(x)_{ij} = z_i \bar{z}_j. \tag{5.13}$$

Similarly,

$$P^{(-1)}(x)_{ij} = \bar{z}_i z_j. \tag{5.14}$$

Inspection shows that z and $\epsilon \bar{z} = (\epsilon_{ij} \bar{z}_j)$ are eigenvectors of $P^{(1)}(x)$ with eigenvalues 1 and 0, whereas $\bar{z}$ and ϵz are those of $P^{(-1)}(x)$ with the same eigenvalues.

Previous remarks on the impossibility of diagonalizing $P^{(k)}(x)$ using a unitary $U(x)$ for all x do not contradict the existence of these eigenvectors: their domain is not S^2, but S^3.

Just as $P^{(\pm 1)}$, $P'^{(k)}$ has eigenvectors $v_k, \epsilon \bar{v}_k$ for $k > 0$, and $\bar{v}_{|k|}, \epsilon v_{-|k|}$ for $k < 0$.

As $P^{(k)}$ is $2^{|k|} \times 2^{|k|}$, let us embed $P'^{(k)}$ inside a $2^{|k|} \times 2^{|k|}$ matrix $\mathcal{P}'^{(k)}$ in the manner described above.

Let us first assume that $k > 0$.

Let $\xi^{(k)}(j)$ be orthonormal eigenvectors of $P^{(k)}$ constructed as follows: For $\xi^{(k)}(1)$, we set

$$\xi^{(k)}(1) = \begin{matrix} z & \otimes & z & \cdots & \otimes & z \\ 1 & & 2 & & & k \end{matrix} \tag{5.15}$$

The integers $1, 2, \cdots, k$ below z's label the vector space $\mathbb{C}^2$ which contains the z above it: the z above j belongs to the $\mathbb{C}^2$ of the j-th slot in the tensor product $\mathbb{C}^2 \otimes \mathbb{C}^2 \otimes \cdots \otimes \mathbb{C}^2 = \mathbb{C}^{2^k}$.

The next set of vectors $\xi^{(k)}(j)$ $(j = 2, \cdots, k+1)$ is obtained by replacing z above j by $\epsilon \bar{z}$ and not touching the remaining z's. We say we have "flipped" one z at a time to get these vectors.

Next we flip 2 z's at a time: there are $_kC_2$ of these.

We proceed in this manner, flipping 3,4, etc z's. When all are flipped, we get the vector

$$\xi^{(k)}(2^k) = \epsilon \bar{z} \otimes \epsilon \bar{z} \otimes \cdots \otimes \epsilon \bar{z}. \tag{5.16}$$

The following is important: a basis vector after j flips has the property

$$\xi^{(k)}(l) \rightarrow e^{i(k-2j)\theta} \xi^{(k)}(l), \quad \text{when} \quad z \rightarrow e^{i\theta} z. \tag{5.17}$$

Our task is to find an orthonormal basis $\eta^{(k)}(l)$ where $\eta^{(k)}(1)$ is the following eigenvector of $\mathcal{P}'^{(k)}(x)$ with eigenvalue 1,

$$\eta^{(k)}(1) = (v_k, \vec{0}),$$
$$\mathcal{P}'^{(k)}(x)\eta^{(k)}(1) = \eta^{(k)}(1). \tag{5.18}$$

Then the rest are in the null space of $\mathcal{P}'^{(k)}(x)$:

$$\mathcal{P}'^{(k)}(x)\eta^{(k)}(j) = 0, \quad j \neq 1. \tag{5.19}$$

We require in addition that $\eta^{(k)}(l)$ transforms in exactly the same manner as $\xi^{(k)}(l)$:

$$\eta^{(k)}(l) \rightarrow e^{i(k-2j)\theta} \eta^{(k)}(l), \quad \text{when} \quad z \rightarrow e^{i\theta} z. \tag{5.20}$$

Then the operator

$$\hat{U}(z) = \sum_l \xi^{(k)}(l) \otimes \bar{\eta}^{(k)}(l) \tag{5.21}$$

is unitary,

$$\hat{U}(z)^\dagger \hat{U}(z) = \mathbf{1}, \tag{5.22}$$

and invariant under $z \to ze^{i\theta}$:

$$\hat{U}(ze^{i\theta}) = \hat{U}(z). \tag{5.23}$$

Hence we can write

$$\hat{U}(z) = U(x) \tag{5.24}$$

and U provides the equivalence between $P'^{(k)}$ and $P^{(k)}$:

$$U\mathcal{P}'^{(k)}U^\dagger = \mathcal{P}^{(k)} \tag{5.25}$$

There are indeed such orthonormal vectors. $\eta^{(k)}(1)$ clearly has the required property. As for the rest, we show how to find them for $k = 2$ and 3. The general construction is similar.

If $k = -|k| < 0$, the same considerations apply after changing z to $\epsilon\bar{z}$ in $P^{(k)}$ and v_k to $\bar{v}_{|k|}$ in $P'^{(k)}$.

k=2

In this case, $\mathbb{C}^{2^k} = \mathbb{C}^4$. The basis is

$$\eta^{(2)}(1) = \begin{pmatrix} v_2 \\ 0 \\ 0 \end{pmatrix}, \quad \eta^{(2)}(2) = \begin{pmatrix} 0 \\ 0 \\ v_2 \end{pmatrix},$$

$$\eta^{(2)}(3) = \begin{pmatrix} 0 \\ 0 \\ \epsilon\bar{v}_2 \end{pmatrix}, \quad \eta^{(2)}(4) = \begin{pmatrix} \epsilon\bar{v}_2 \\ 0 \\ 0 \end{pmatrix}. \tag{5.26}$$

k=3

Now $\mathbb{C}^{2^k} = \mathbb{C}^8$. The basis is

$$
\eta^{(3)}(1) = \begin{pmatrix} v_3 \\ 0 \\ 0 \\ 0 \\ 0 \\ 0 \\ 0 \end{pmatrix}, \quad
\eta^{(3)}(2) = \begin{pmatrix} 0 \\ 0 \\ v_3 \\ 0 \\ 0 \\ 0 \\ 0 \end{pmatrix}, \quad
\eta^{(3)}(3) = \begin{pmatrix} 0 \\ 0 \\ 0 \\ 0 \\ v_3 \\ 0 \\ 0 \end{pmatrix},
$$

$$
\eta^{(3)}(4) = \begin{pmatrix} 0 \\ 0 \\ 0 \\ 0 \\ 0 \\ 0 \\ v_3 \end{pmatrix}, \quad
\eta^{(3)}(5) = \begin{pmatrix} 0 \\ 0 \\ \epsilon\bar{v}_3 \\ 0 \\ 0 \\ 0 \\ 0 \end{pmatrix}, \quad
\eta^{(3)}(6) = \begin{pmatrix} 0 \\ 0 \\ 0 \\ 0 \\ \epsilon\bar{v}_3 \\ 0 \\ 0 \end{pmatrix},
$$

$$
\eta^{(3)}(7) = \begin{pmatrix} 0 \\ 0 \\ 0 \\ 0 \\ 0 \\ 0 \\ \epsilon\bar{v}_3 \end{pmatrix}, \quad
\eta^{(3)}(8) = \begin{pmatrix} \epsilon\bar{v}_3 \\ 0 \\ 0 \\ 0 \\ 0 \\ 0 \\ 0 \end{pmatrix}. \tag{5.27}
$$

In this manner, we can always construct $\eta^{(k)}(j)$.

5.4 Projective Modules on the Fuzzy Sphere

We want to construct the analogues of $P^{(k)}$ and $P'^{(k)}$ for the fuzzy sphere. They give us the monopoles and instantons of S_F^2. Let us consider $P^{(k)}$ first, and denote the corresponding projectors as $P_F^{(k)}$.

5.4.1 *Fuzzy Monopoles and Projectors $P_F^{(k)}$*

We begin by illustrating the ideas for $k = 1$.

On $\mathbb{C}^2$, the spin $\frac{1}{2}$ representation of $SU(2)$ acts with generators $\frac{\tau_i}{2}$. On S_F^2, the spin ℓ representation of $SU(2)$ acts with generators L_i^L. Let $P_F^{(1)}$ be the projector coupling ℓ and $\frac{1}{2}$ to $\ell + \frac{1}{2}$. Consider the projective module

$P_F^{(1)}(S_F^2 \otimes \mathbb{C}^2)$. On this module,

$$(\vec{L}^L + \frac{\vec{\tau}}{2})^2 = (\ell + \frac{1}{2})(\ell + \frac{3}{2}), \tag{5.28}$$

or

$$\frac{\vec{L}^L}{\ell} \cdot \vec{\tau} = 1. \tag{5.29}$$

Passing to the limit $\ell \to \infty$, this becomes $\hat{x} \cdot \vec{\tau} = 1$, so $P_F^{(1)} \to P^{(1)}$ as $\ell \to \infty$.

We can find $P_F^{(1)}$ explicitly.

$$-2P_F^{(1)} - 1 \equiv \Gamma^L = \frac{\vec{\tau} \cdot \vec{L}^L + \frac{1}{2}}{\ell + \frac{1}{2}}. \tag{5.30}$$

Γ^L is an involution,

$$(\Gamma^L)^2 = 1 \tag{5.31}$$

and will turn up in the theory of fuzzy Dirac operators and the Ginsparg-Wilson system (see chapter 8). It is the chirality operator of the Watamuras' [69].

An important feature of $P_F^{(1)}(S_F^2 \otimes \mathbb{C}^2)$ is that it is still an $SU(2)$-bimodule. On the right, L_i^R act as before. On the left, L_i^L do not, but $L_i^L + \frac{\tau_i}{2}$ do as they commute with $P_F^{(1)}$.

This addition of $\frac{\vec{\tau}}{2}$ to $\vec{L}^L$ stands here for the phenomenon of "mixing of spin and isospin" in the "t Hooft-Polyakov monopole theory [70].

But $P_F^{(1)}(S_F^2 \otimes \mathbb{C}^2)$ is not a free S_F^2-module as it does not have a basis $\{e_i = (e_{i1}, e_{i2}) : e_{i,j} \in S_F^2\}$. That is because if $\alpha = (\alpha_1, \alpha_2) \in S_F^2 \otimes \mathbb{C}^2, \alpha_i \in S_F^2$ the projector $P_F^{(1)}$ mixes up the rows of α_i.

For $k = -1$, the projector $P_F^{(-1)}$ couples ℓ and $1/2$ to $\ell - 1/2$. It is just $1 - P_F^{(1)}$.

The construction for any k is similar. For $k = |k|$, we consider $\mathbb{C}^{2^k} = \mathbb{C}^2 \otimes \mathbb{C}^2 \cdots \otimes \mathbb{C}^2$. On this, the $SU(2)$ acts on each $\mathbb{C}^2$, the generators for the jth slot being $\tau_i^{(j)}/2 \equiv \mathbf{1} \otimes \cdots \otimes \frac{\tau_i}{2} \otimes \cdots \otimes \mathbf{1}$, the $\frac{\tau_i}{2}$ being in the jth slot. Let $P_F^{(k)}$ be the projector coupling ℓ and all the spin $\frac{1}{2}$'s to the maximum value $\ell + \frac{k}{2}$. The projective module is $P_F^{(k)}(S_F^2 \otimes \mathbb{C}^{2^k})$.

For $k = -|k|$, $P_F^{(k)}$ couples ℓ and the spins to the least value $\ell - \frac{|k|}{2}$.

We can show that $\frac{(\tau^{(j)} \cdot L^L)}{\ell}$ tends to $+1$ for $k > 0$ and -1 for $k < 0$ on these modules, so that the $\tau^{(j)} \cdot \hat{x}$ have the correct values in the limit. Thus consider for example $k > 0$. As all angular momenta are coupled to the maximum possible value, every pair must also be so coupled. So on this module $(\vec{L}^L + \frac{\vec{\tau}^{(j)}}{2})^2 = (\ell + \frac{1}{2})(\ell + \frac{3}{2})$ and the result follows as for $k = 1$.

Similar considerations apply for $k < 0$.

For higher k, we can also proceed in a different manner. If $k = |k|$, $SU(2)$ acts on $\mathbb{C}^{k+1}$ by angular momentum $\frac{k}{2}$ representation. Hence there is the projector $P'^{(k)}$ coupling the left ℓ and $\frac{k}{2}$ to $\ell + \frac{k}{2}$. The projective module is then $P'^{(k)}(S_F^2 \otimes \mathbb{C}^{k+1})$.

For $k < 0$ we can couple ℓ and $|k|$ to $\ell - \frac{|k|}{2}$ instead (we assume $\ell > \frac{|k|}{2}$).

$P'^{(k)}$ and $P^{(k)}$ are equivalent in the sense discussed earlier. We can in fact exhibit the two modules so that they look the same: diagonalize the angular momentum $(\vec{L}^L + \sum_j \frac{\vec{\tau}^{(j)}}{2})^2$ and its third component on $P_F^{(k)}(S_F^2 \otimes \mathbb{C}^{2^k})$. Their right angular momenta being both ℓ, their equivalence (in any sense!) is clear.

For reasons indicated above, none of these S_F^2-modules are free.

5.4.2 The Fuzzy Module for the Tangent Bundle and the Fuzzy Complex Structure

The projectors for $k = 2$ are of particular interest as they can be interpreted as fuzzy sections of the tangent bundle.

To see this, let us begin with the commutative algebra $\mathcal{A} = C^\infty(S^2)$ and the module $\mathcal{A}^3 = C^\infty(S^2) \otimes \mathbb{C}^3$. In this case, $SU(2)$ acts on $\mathbb{C}^3$ with the spin 1 generators $\theta(\alpha)$ where

$$\theta(\alpha)_{ij} = -i\epsilon_{\alpha ij}. \tag{5.32}$$

Consider

$$\theta(\alpha)\hat{x}_\alpha \equiv \theta \cdot \hat{x}. \tag{5.33}$$

Its eigenvalues at each x are $\pm 1, 0$. Let $P^{(T)}$ be the projector to the subspace $(\theta \cdot \hat{x})^2 = 1$:

$$P^{(T)} = (\theta \cdot \hat{x})^2. \tag{5.34}$$

Any vector in the module $P^{(T)}\mathcal{A}^3$ can be written as $\xi^+ + \xi^-$ where $\theta \cdot \hat{x}\xi^\pm = \pm\xi^\pm$, that is $-i\epsilon_{\alpha ij}x_\alpha\xi_j^\pm(x) = \pm\xi_i^\pm(x)$. It follows from the antisymmetry of $\epsilon_{\alpha ij}$ that $x_i\xi_i^\pm(x) = 0$ or that $\xi^\pm(x)$ are tangent to S^2 at x. The $\xi^\pm$ give sections of the (complexified) tangent bundle TS^2.

A smooth split for all x of $TS^2(x)$ into two subspaces $TS_\pm^2(x)$ gives a complex structure J on TS^2. $J(x)$ is $\pm i\mathbf{1}$ on $TS_\pm^2(x)$. Thus a complex structure on TS^2 is defined by the decomposition

$$TS^2 = TS_+^2 \oplus TS_-^2,$$

$$J|_{TS_\pm^2} = \pm i\mathbf{1}. \tag{5.35}$$

Now $P^{(T)}$ is the sum of projectors which give eigenspaces of $\theta \cdot \hat{x}$ for eigenvalues ± 1:

$$P^{(T)} = P_+^{(T)} + P_-^{(T)},$$
$$P_\pm^{(T)} = \frac{\theta \cdot \hat{x}(\theta \cdot \hat{x} \pm 1)}{2}. \tag{5.36}$$

With

$$JP_\pm^{(T)} = \pm i P_\pm^{(T)} \tag{5.37}$$

we get the required decomposition of $P^{(1)}\mathcal{A}^3$ for a complex structure:

$$P^{(T)}\mathcal{A}^3 = P_+^{(T)}\mathcal{A}^3 \oplus P_-^{(T)}\mathcal{A}^3. \tag{5.38}$$

Fuzzification of these structures is easy and elegant.

Instead of working with $S_F^2 \otimes \mathbb{C}^2$ we work with $S_F^2 \otimes \mathbb{C}^3$. The projector $P_F^{(T)}$ we thereby obtain is the fuzzy version of P^T. We can show this as follows.

Let $P_F^{(T,\pm)}$ be the projectors coupling L_α^L and $\theta(\alpha)$ to the values $\ell \pm 1$. Then

$$P_F^{(T)} = P_F^{(T,+)} + P_F^{(T,-)}. \tag{5.39}$$

On the module $P_F^{(T,+)}(S_F^2 \otimes \mathbb{C}^3)$,

$$[L_\alpha^L + \theta(\alpha)]^2 = (\ell+1)(\ell+2) \tag{5.40}$$

or

$$\frac{L_\alpha^L \theta(\alpha)}{\ell} = 1. \tag{5.41}$$

On the module $P_F^{(T,-)}(S_F^2 \otimes \mathbb{C}^3)$,

$$(L_\alpha^L + \theta(\alpha))^2 = -1 - \frac{1}{\ell} \tag{5.42}$$

Thus as $\ell \to \infty$

$$\frac{L_\alpha^L \theta(\alpha)}{\ell} \to \pm 1 \quad \text{on} \quad P_F^{(T,\pm)}(S_F^2 \otimes \mathbb{C}^3). \tag{5.43}$$

As the left hand side tends to $\theta(\alpha)\hat{x}_\alpha$ as $\ell \to \infty$, we have that $P_F^{(T)}(S_F^2 \otimes \mathbb{C}^3)$ defines the fuzzy tangent bundle and its decomposition $P_F^{(T,+)}(S_F^2 \otimes \mathbb{C}^3) \oplus P_F^{(T,-)}(S_F^2 \otimes \mathbb{C}^3)$ defines the fuzzy complex structure. The corresponding J, call it J_F, is $\pm i$ on $P_F^{(T,\pm)}(S_F^2 \otimes \mathbb{C}^3)$.

Chapter 6

Fuzzy Nonlinear Sigma Models

6.1 Introduction

In space-time dimensions larger than 2, whenever a global symmetry G is spontaneously broken to a subgroup H, and G and H are Lie groups, there are massless Nambu-Goldstone modes with values in the coset space G/H. Being massless, they dominate low energy physics as is the case with pions in strong interactions and phonons in crystals. Their theoretical description contains new concepts because G/H is not a vector space.

Such G/H models have been studied extensively in $2 - d$ physics, even though in that case there is no spontaneous breaking of continuous symmetries. A reason is that they are often tractable nonperturbatively in the two-dimensional context, and so can be used to test ideas suspected to be true in higher dimensions. A certain amount of numerical work has also been done on such $2 - d$ models to control conjectures and develop ideas, their discrete versions having been formulated for this purpose.

This chapter develops discrete fuzzy approximations to G/H models. We focus on two-dimensional Euclidean quantum field theories with target space $G/H = SU(N + 1)/U(N) = \mathbb{C}P^N$. The novelty of this approach is that it is based on fuzzy physics [3] and non-commutative geometry [34–38]. Although fuzzy physics has striking elegance because it preserves the symmetries of the continuum and because techniques of non-commutative geometry give us powerful tools to describe continuum topological features, still its numerical efficiency has not been fully tested. This chapter approaches σ-models with this in mind, the idea being to write fuzzy G/H models in a form adapted to numerical work.

This is not the only approach on fuzzy G/H. In [81], a particular description based on projectors and their orbits was discretized. We shall refine that work considerably in this paper. Also in the continuum there is another way to approach G/H, namely as gauge theories with gauge invariance under H and global symmetry under G [72]. This approach is extended here to fuzzy physics. Such a fuzzy gauge theory involves the decomposition of projectors in terms of partial isometries [68] and brings new ideas into this field. It is also very pretty. It is equivalent to the projector method as we shall also see.

Related work on fuzzy G/H model and their solitons is due to Govindarajan and Harikumar [73]. A different treatment, based on the Holstein-Primakoff realization of the $SU(2)$ algebra, has been given in [74]. A more general approach to these models on noncommutative spaces was proposed in [75].

The first two sections describe the standard $\mathbb{C}P^1$-models on S^2. In section 2, we discuss it using projectors, while in section 3, we reformulate the discussion in such a manner that transition to fuzzy spaces is simple. Sections 4 and 5 adapt the previous sections to fuzzy spaces.

Long ago, general G/H-models on S^2 were written as gauge theories [72]. Unfortunately their fuzzification for generic G and H eludes us. Generalization of the considerations here to the case where $S^2 \simeq \mathbb{C}P^1$ is replaced with $\mathbb{C}P^N$, or more generally Grassmannians and flag manifolds associated with $(N+1) \times (N+1)$ projectors of rank $\leq (N+1)/2$, is easy as we briefly show in the concluding section 6. But extension to higher ranks remains a problem.

6.2 CP^1 Models and Projectors

Let the unit vector $x = (x_1, x_2, x_3) \in \mathbb{R}^3$ describe a point of S^2. The field n in the $\mathbb{C}P^1$-model is a map from S^2 to S^2:

$$n = (n_1, n_2, n_3) : x \rightarrow n(x) \in \mathbb{R}^3, \quad n(x) \cdot n(x) := \sum_a n_a(x)^2 = 1 . \quad (6.1)$$

These maps n are classified by their winding number $\kappa \in \mathbb{Z}$:

$$\kappa = \frac{1}{8\pi} \int_{S^2} \epsilon_{abc} \, n_a(x) \, dn_b(x) \, dn_c(x) . \quad (6.2)$$

That κ is the winding of the map can be seen taking spherical coordinates (Θ, Φ) on the target sphere $(n^2 = 1)$ and using the identity $\sin\Theta d\Theta \, d\Phi = \frac{1}{2}\epsilon_{abc} n_a dn_b \, dn_c$. We omit wedge symbols in products of forms.

We can think of n as the field at a fixed time t on a $(2+1)$-dimensional manifold where the spatial slice is S^2. In that case, it can describe a field of spins, and the fields with $\kappa \neq 0$ describe solitonic sectors. We can also think of it as a field on Euclidean space-time S^2. In that case, the fields with $\kappa \neq 0$ describe instantonic sectors.

Let τ_a be the Pauli matrices. Then each $n(x)$ is associated with the projector

$$P(x) = \frac{1}{2}(1 + \vec{\tau} \cdot \vec{n}(x)) . \qquad (6.3)$$

Conversely, given a 2×2 projector $P(x)$ of rank 1, we can write

$$P(x) = \frac{1}{2}(\alpha_0(x) + \vec{\tau} \cdot \vec{\alpha}(x)) . \qquad (6.4)$$

Using $\operatorname{Tr} P(x) = 1$, $P(x)^2 = P(x)$ and $P(x)^\dagger = P(x)$, we get

$$\alpha_0(x) = 1, \quad \vec{\alpha}(x) \cdot \vec{\alpha}(x) = 1, \quad \alpha_a^*(x) = \alpha_a(x) . \qquad (6.5)$$

Thus $\mathbb{C}P^1$-fields on S^2 can be described either by P or by $n_a = \operatorname{Tr}(\tau_a P)$ [76].

In terms of P, κ is

$$\kappa = \frac{1}{2\pi i} \int_{S^2} \operatorname{Tr} P \, (dP) \, (dP) . \qquad (6.6)$$

There is a family of projectors, called Bott projectors [77, 78], which play a central role in our approach. Let

$$z = (z_1, z_2), \quad |z|^2 := |z_1|^2 + |z_2|^2 = 1 . \qquad (6.7)$$

The z's are points on S^3. We can write $x \in S^2$ in terms of z:

$$x_i(z) = z^\dagger \tau_i z . \qquad (6.8)$$

The Bott projectors are

$$P_\kappa(x) = v_\kappa(z) v_\kappa^\dagger(z), \quad v_\kappa(z) = \begin{bmatrix} z_1^\kappa \\ z_2^\kappa \end{bmatrix} \frac{1}{\sqrt{Z_\kappa}} \quad \text{if } \kappa \geq 0 ,$$

$$Z_k \equiv |z_1|^{2|\kappa|} + |z_2|^{2|\kappa|} ,$$

$$v_\kappa(z) = \begin{bmatrix} z_1^{*|\kappa|} \\ z_2^{*|\kappa|} \end{bmatrix} \frac{1}{\sqrt{Z_\kappa}} \quad \text{if } \kappa < 0 . \qquad (6.9)$$

The field $n^{(\kappa)}$ associated with P_κ is given by

$$n_a^{(\kappa)}(x) = \operatorname{Tr} \tau_a P_\kappa(x) = v_\kappa^\dagger(z) \tau_a v_\kappa(z) . \qquad (6.10)$$

Under the phase change $z \to z e^{i\theta}$, $v_\kappa(z)$ changes, $v_\kappa(z) \to v_\kappa(z) e^{i\kappa\theta}$, whereas x is invariant. As this phase cancels in $v_\kappa(z) v_\kappa^\dagger(z)$, P_κ is a function of x as written.

The κ that appears in eqs.(6.9)(6.10) is the winding number as the explicit calculation of section 3 will show. But there is also the following argument.

In the map $z \rightarrow v_\kappa(z)$, for $\kappa = 0$, all of S^3 and S^2 get mapped to a point, giving zero winding number. So, consider $\kappa > 0$. Then the points

$$\left(z_1 e^{i\frac{2\pi}{\kappa}(l+m)}, z_2 e^{i\frac{2\pi}{\kappa}m} \right), \quad l, m \in \{0, 1, .., \kappa - 1\} \tag{6.11}$$

have the same image. But the overall phase $e^{i\frac{2\pi}{\kappa}m}$ of z cancels out in x. Thus, generically κ points of S^2 (labeled by l) have the same projector $P_\kappa(x)$, giving winding number κ. As for $\kappa < 0$, we get $|\kappa|$ points of S^2 mapped to the same $P_\kappa(x)$. But because of the complex conjugation in eq.(6.9), there is an orientation-reversal in the map giving $-|\kappa| = \kappa$ as winding numbers. One way to see this is to use

$$P_{-|\kappa|}(x) = P_{|\kappa|}(x)^T \tag{6.12}$$

Substituting this in (6.6), we can see that $P_{\pm|\kappa|}$ have opposite winding numbers.

The general projector $\mathcal{P}_\kappa(x)$ is the gauge transform of $P_\kappa(x)$:

$$\mathcal{P}_\kappa(x) = U(x)P_\kappa(x)U(x)^\dagger, \tag{6.13}$$

where $U(x)$ is a unitary 2×2 matrix. Its $n^{(\kappa)}$ is also given by (6.10), with P_κ replaced by $\mathcal{P}_\kappa$. The winding number is unaffected by the gauge transformation. That is because U is a map from S^2 to $U(2)$ and all such maps can be deformed to identity since $\pi_2(U(2)) = \{\text{identity } e\}$.

The identity

$$\mathcal{P}_\kappa(d\mathcal{P}_\kappa) = (d\mathcal{P}_\kappa)(\mathbb{1} - \mathcal{P}_\kappa) \tag{6.14}$$

which follows from $\mathcal{P}_\kappa^2 = \mathcal{P}_\kappa$, is valuable when working with projectors.

The soliton described by P_κ has the action (below) peaked at the north pole $x_3 = 1$ or $\frac{x_1 + ix_2}{1+x_3} = 0$ and a fixed width and shape. The solitons with energy density peaked at $\frac{x_1 + ix_2}{1+x_3} = \eta$ and variable width and shape are given by the projectors

$$P_\kappa(x, \eta, \lambda) = v_\kappa(z, \eta, \lambda)v_\kappa(z, \eta, \lambda)^\dagger,$$

$$v_\kappa(z, \eta, \lambda) = \begin{pmatrix} \lambda z_1^\kappa \\ z_2^\kappa - \eta z_1^\kappa \end{pmatrix} \frac{1}{(|\lambda z_1|^{2\kappa} + |z_2^\kappa - \eta z_1^\kappa|^2)^{\frac{1}{2}}}. \tag{6.15}$$

In order to find the explicit form of the unitary transformation in (6.13) corresponding to the choice (6.15) we proceed as follows. There is a unit vector $w_\kappa(.)$ perpendicular to $v_\kappa(.)$ (The dot represents z or z, η, λ). An explicit realization of w_κ is given by

$$w_{\kappa,\alpha} = i(\tau_2)_{\alpha\beta} v^*_{\kappa,\beta} := \epsilon_{\alpha\beta} v^*_{\kappa,\beta} \tag{6.16}$$

from which it is easy to see that $w^\dagger_\kappa(.)v_\kappa(.) = 0 = v^\dagger_\kappa(.)w_\kappa(.)$. $w_\kappa(.)$ also fulfills

$$w_\kappa(.)w^\dagger_\kappa(.) = \mathbb{1} - P_\kappa(.) , \quad w^\dagger_\kappa(.)w_\kappa(.) = 1 . \tag{6.17}$$

Let us consider

$$V(x) = v_\kappa(z, \eta, \lambda)v^\dagger_\kappa(z) , \quad W(x) = w_\kappa(z, \eta, \lambda)w^\dagger_\kappa(z) . \tag{6.18}$$

From these definitions it is easy to verify that

$$U(x) = V(x) + W(x) \tag{6.19}$$

does the required job for $\kappa > 0$.

We call the field associated with $P_\kappa(., \eta, \lambda)$ as $n^{(\kappa)}(., \eta, \lambda)$:

$$\vec{n}^{(\kappa)}(x, \lambda, \eta) = v_\kappa(z, \eta, \lambda)^\dagger \vec{\tau} v_\kappa(z, \eta, \lambda) . \tag{6.20}$$

We can use $v_\kappa(z, \eta, \lambda) = v_{|\kappa|}(\bar{z}, \eta, \lambda)$ and $w_\kappa(z, \eta, \lambda) = w_{|\kappa|}(\bar{z}, \eta, \lambda)$ to write the solitons for $\kappa < 0$.

6.3 An Action

Let $\mathcal{L}_i = -i(x \wedge \nabla)_i$ be the angular momentum operator. Then a Euclidean action in the κ-th topological sector (or a static Hamiltonian in the $(2+1)$ picture) is

$$S_\kappa = -\frac{c}{2} \int_{S^2} d\Omega \, (\mathcal{L}_i n_b^{(\kappa)})(\mathcal{L}_i n_b^{(\kappa)}) , \quad c = \text{ a positive constant}, \tag{6.21}$$

where $d\Omega$ is the S^2 volume form $d\cos\theta \, d\varphi$. We can also write

$$S_\kappa = -c \int_{S^2} d\Omega \, \text{Tr} \, (\mathcal{L}_i P_\kappa)(\mathcal{L}_i P_\kappa) . \tag{6.22}$$

The following identities, based on (6.14), are also useful:

$$\text{Tr} \, P_\kappa(\mathcal{L}_i P_\kappa)^2 = \text{Tr} \, (\mathcal{L}_i P_\kappa)(\mathbb{1} - P_\kappa)(\mathcal{L}_i P_\kappa) =$$
$$\text{Tr}(\mathbb{1} - P_\kappa)(\mathcal{L}_i P_\kappa)^2 = \frac{1}{2} \text{Tr}(\mathcal{L}_i P_\kappa)^2 . \tag{6.23}$$

Hence

$$S_\kappa = -2c \int_{S^2} d\Omega \, \text{Tr} \, P_\kappa \, \mathcal{L}_i P_\kappa \, \mathcal{L}_i P_\kappa \tag{6.24}$$

The Euclidean functional integral for the actions S_κ is

$$Z(\psi) = \sum_\kappa e^{i\kappa\psi} \int \mathcal{D}\mathcal{P}_\kappa e^{-S_\kappa} \qquad (6.25)$$

where the angle ψ is induced by the instanton sectors as in QCD.

Using the identity $dP = -\epsilon_{ijk}\, dx_i\, x_j\, i\mathcal{L}_k P$, we can rewrite the definition (6.2) or (6.6) of the winding number as

$$\kappa = \frac{1}{8\pi} \int_{S^2} d\Omega\, \epsilon_{ijk} x_i\, \epsilon_{abc} n_a^{(\kappa)}\, i\mathcal{L}_j n_b^{(\kappa)}\, i\mathcal{L}_k n_c^{(\kappa)} \qquad (6.26)$$

$$= \frac{1}{2\pi i} \int_{S^2} d\Omega\, \mathrm{Tr}\, \mathcal{P}_\kappa\, \epsilon_{ijk}\, x_i\, i\mathcal{L}_j \mathcal{P}_\kappa\, i\mathcal{L}_k \mathcal{P}_\kappa \,. \qquad (6.27)$$

The Belavin-Polyakov bound [79]

$$S_\kappa \geq 4\pi\, c\, |\kappa| \qquad (6.28)$$

follows from (6.26) on integration of

$$(i\mathcal{L}_i n_a^{(\kappa)} \pm \epsilon_{ijk} x_j\, \epsilon_{abc}\, n_b^{(\kappa)}\, i\mathcal{L}_k n_c^{(\kappa)})^2 \geq 0 \,, \qquad (6.29)$$

or from (6.27) on integration of

$$\mathrm{Tr}\left(\mathcal{P}_\kappa(i\mathcal{L}_i\mathcal{P}_\kappa) \pm i\epsilon_{ijk}\, x_j \mathcal{P}_\kappa(i\mathcal{L}_k\mathcal{P}_\kappa)\right)^\dagger \left(\mathcal{P}_\kappa(i\mathcal{L}_i\mathcal{P}_\kappa) \pm i\epsilon_{ij'k'}\, x_{j'}\mathcal{P}_\kappa(i\mathcal{L}_{k'}\mathcal{P}_\kappa)\right) \geq 0 \,. \qquad (6.30)$$

From this last form it is easy to rederive the bound in a way better adapted to fuzzification. Using Pauli matrices $\{\sigma_i\}$ we first rewrite (6.24) and (6.27) as

$$S_\kappa = c \int_{S^2} d\Omega\, \mathrm{Tr}\, \mathcal{P}_\kappa(i\sigma \cdot \mathcal{L}\mathcal{P}_\kappa)(i\sigma \cdot \mathcal{L}\mathcal{P}_\kappa) \,,$$

$$\kappa = \frac{-1}{4\pi} \int_{S^2} d\Omega\, \mathrm{Tr}\, (\sigma \cdot x\, \mathcal{P}_k(i\sigma \cdot \mathcal{L}\mathcal{P}_k)(i\sigma \cdot \mathcal{L}\mathcal{P}_k)) \,. \qquad (6.31)$$

The trace is now over $\mathbb{C}^2 \times \mathbb{C}^2 = \mathbb{C}^4$, where τ_a acts on the first $\mathbb{C}^2$ and σ_i on the second $\mathbb{C}^2$ (so they are really $\tau_a \otimes \mathbb{1}$ and $\mathbb{1} \otimes \sigma_i$) Then, with $\epsilon_1, \epsilon_2 = \pm 1$,

$$\frac{1 + \epsilon_2\tau \cdot n^{(\kappa)}}{2}\sigma_i\left((i\mathcal{L}_i\mathcal{P}_\kappa) + \epsilon_1 i\epsilon_{ijk}\, x_j(i\mathcal{L}_k\mathcal{P}_\kappa)\right) =$$

$$(1 + \epsilon_1\sigma \cdot x)\frac{1 + \epsilon_2\tau \cdot n^{(\kappa)}}{2}(i\sigma \cdot \mathcal{L}\mathcal{P}_\kappa) \,, \qquad (6.32)$$

since $x \cdot \mathcal{L} = 0$. The inequality (6.30) is equivalent to

$$\mathrm{Tr}\left(\left[\frac{1 + \epsilon_1\sigma \cdot x}{2}\frac{1 + \epsilon_2\tau \cdot n^{(\kappa)}}{2}(i\sigma \cdot \mathcal{L}\mathcal{P}_\kappa)\right]^\dagger \times \right.$$

$$\left.\left[\frac{1 + \epsilon_1\sigma \cdot x}{2}\frac{1 + \epsilon_2\tau \cdot n^{(\kappa)}}{2}(i\sigma \cdot \mathcal{L}\mathcal{P}_\kappa)\right]\right) \geq 0 \,, \qquad (6.33)$$

from which (6.28) follows by integration.

6.4 $\mathbb{C}P^1$-Models and Partial Isometries

If $\mathcal{P}(x)$ is a rank 1 projector at each x, we can find its normalized eigenvector $u(z)$:

$$\mathcal{P}(x)u(z) = u(z), \quad u^\dagger(z)u(z) = 1 . \tag{6.34}$$

Then

$$\mathcal{P}(x) = u(z)u^\dagger(z) . \tag{6.35}$$

If $\mathcal{P} = \mathcal{P}_\kappa$, an example of u is v_κ. u can be a function of z, changing by a phase under $z \to ze^{i\theta}$. Still, $\mathcal{P}$ depends only on x.

We can regard $u(z)^\dagger$ (or a slight generalization of it) as an example of a partial isometry [68] in the algebra $\mathcal{A} = C^\infty(S^3) \otimes_\mathbb{C} Mat_{2\times2}(\mathbb{C})$ of 2×2 matrices with coefficients in $C^\infty(S^3)$. A partial isometry in a $*-$algebra A is an element $\mathcal{U}^\dagger \in A$ such that $\mathcal{U}\mathcal{U}^\dagger$ is a projector. $\mathcal{U}\mathcal{U}^\dagger$ is the *support projector* of $\mathcal{U}^\dagger$. It is an isometry if $\mathcal{U}^\dagger\mathcal{U} = \mathbb{1}$. With

$$\mathcal{U} = \begin{pmatrix} u_1 & 0 \\ u_2 & 0 \end{pmatrix} \in \mathcal{A}, \tag{6.36}$$

we have

$$\mathcal{P} = \mathcal{U}\mathcal{U}^\dagger \tag{6.37}$$

so that $\mathcal{U}^\dagger$ is a partial isometry.

We will be free with language and also call $u^\dagger$ as a partial isometry.

The partial isometry for P_κ is $v_\kappa^\dagger$.

Now consider the one-form

$$A_\kappa = v_\kappa^\dagger dv_\kappa . \tag{6.38}$$

Under $z_i \to z_i e^{i\theta(x)}$, A_κ transforms like a connection:

$$A_\kappa \to A_\kappa + i\kappa \, d\theta \tag{6.39}$$

(A_κ are connections for $U(1)$ bundles on S^2 for Chern numbers κ, see later.) Therefore

$$D_\kappa = d + A_\kappa \tag{6.40}$$

is a covariant differential, transforming under $z \to ze^{i\theta}$ as

$$D_\kappa \to e^{i\kappa\theta} D_\kappa e^{-i\kappa\theta} \tag{6.41}$$

and

$$D_\kappa^2 = dA_\kappa \tag{6.42}$$

is its curvature.

Recall that $w(z)$ perpendicular to $v(z)$ for all z was defined in section 6.2. An explicit realization of $w(z)$ is given in (6.16). We have that

$$B_\kappa = w_\kappa^\dagger \, dv_\kappa \quad \text{and} \quad B_\kappa^* = (dv_\kappa^\dagger)w_\kappa = -v_\kappa^\dagger \, dw_\kappa \qquad (6.43)$$

are gauge covariant,

$$B_\kappa(z) \to e^{i\theta(x)} B_\kappa e^{i\theta(x)} \ , \quad B_\kappa(z)^* \to e^{-i\theta(x)} B_\kappa^* e^{-i\theta(x)} \qquad (6.44)$$

under $z \to z e^{i\theta}$.

We can account for $U(x)$ by considering

$$\mathcal{V}_\kappa = U v_\kappa \ , \quad \mathcal{A}_\kappa = \mathcal{V}_\kappa^\dagger \, d\mathcal{V}_\kappa \ , \quad \mathcal{D}_\kappa = d + \mathcal{A}_\kappa \ , \quad \mathcal{D}_\kappa^2 = d\mathcal{A}_\kappa \ ,$$
$$\mathcal{W}_\kappa = (\tau_2 U^* \tau_2) w_\kappa \ , \quad \mathcal{B}_\kappa = \mathcal{W}_\kappa^\dagger \, d\mathcal{V}_\kappa \ . \qquad (6.45)$$

$\mathcal{A}_\kappa$ is still a connection, and the properties (6.44) are not affected by U. $\mathcal{P}_\kappa$ is the support projector of $\mathcal{V}_\kappa^\dagger$, and

$$\mathcal{W}_\kappa \mathcal{W}_\kappa^\dagger = \mathbb{1} - \mathcal{P}_\kappa \ , \quad (\mathbb{1} - \mathcal{P}_\kappa)\mathcal{V}_\kappa = 0 \ . \qquad (6.46)$$

Gauge invariant quantities being functions on S^2, we can contemplate a formulation of the $\mathbb{C}P^1$-model as a gauge theory. Let $\mathcal{J}_i$ be the lift of L_i to angular momentum generators appropriate for functions of z,

$$(e^{i\theta_i \mathcal{J}_i} f)(z) = f(e^{-i\theta_i \tau_i/2} z) \ , \qquad (6.47)$$

and let

$$\mathcal{B}_{\kappa,i} = \mathcal{W}_\kappa^\dagger \mathcal{J}_i \mathcal{V}_\kappa \ . \qquad (6.48)$$

Now, $\mathcal{W}_\kappa \mathcal{B}_{\kappa,i} \mathcal{V}_\kappa^\dagger$ is gauge invariant, and should have an expression in terms of $\mathcal{P}_\kappa$. Indeed it is, in view of (6.46),

$$\mathcal{W}_\kappa \mathcal{B}_{\kappa,i} \mathcal{V}_\kappa^\dagger = \mathcal{W}_\kappa \mathcal{W}_\kappa^\dagger (\mathcal{J}_i \mathcal{V}_\kappa) \mathcal{V}_\kappa^\dagger =$$
$$(\mathbb{1} - \mathcal{P}_\kappa)\mathcal{J}_i(\mathcal{V}_\kappa \mathcal{V}_\kappa^\dagger) = (\mathbb{1} - \mathcal{P}_\kappa)(\mathcal{L}_i \mathcal{P}_\kappa) = (\mathcal{L}_i \mathcal{P}_\kappa)\mathcal{P}_\kappa \ . \qquad (6.49)$$

Therefore we can write the action (6.22, 6.24) in terms of the $\mathcal{B}_{\kappa,i}$:

$$S_\kappa = -2c \int_{S^2} d\Omega \, \mathrm{Tr} \, \mathcal{P}_\kappa(\mathcal{L}_i \mathcal{P}_\kappa)(\mathcal{L}_i \mathcal{P}_\kappa)$$
$$= 2c \int_{S^2} d\Omega \, \mathrm{Tr} \, ((\mathcal{L}_i \mathcal{P}_\kappa)\mathcal{P}_\kappa)^\dagger ((\mathcal{L}_i \mathcal{P}_\kappa)\mathcal{P}_\kappa)$$
$$= 2c \int_{S^2} d\Omega \, \mathrm{Tr} (\mathcal{W}_\kappa \mathcal{B}_{\kappa,i} \mathcal{V}_\kappa^\dagger)^\dagger (\mathcal{W}_\kappa \mathcal{B}_{\kappa,i} \mathcal{V}_\kappa^\dagger)$$
$$= 2c \int_{S^2} d\Omega \, \mathcal{B}_{\kappa,i}^* \mathcal{B}_{\kappa,i} \ . \qquad (6.50)$$

It is instructive also to write the gauge invariant $(d\mathcal{A}_\kappa)$ in terms of $\mathcal{P}_\kappa$ and relate its integral to the winding number (6.6). The matrix of forms

$$\mathcal{V}_\kappa(d+\mathcal{A}_\kappa)\mathcal{V}_\kappa^\dagger \tag{6.51}$$

is gauge invariant. Here

$$d\mathcal{V}_\kappa^\dagger = (d\mathcal{V}_\kappa^\dagger) + \mathcal{V}_\kappa^\dagger d \tag{6.52}$$

where d in the first term differentiates only $\mathcal{V}_\kappa^\dagger$. Now

$$\mathcal{V}_\kappa(d+\mathcal{V}_\kappa^\dagger(d\mathcal{V}_\kappa))\mathcal{V}_\kappa^\dagger \tag{6.53}$$

and

$$\mathcal{P}_\kappa \, d\mathcal{P}_\kappa = \mathcal{V}_\kappa \mathcal{V}_\kappa^\dagger \, d\,(\mathcal{V}_\kappa \mathcal{V}_\kappa^\dagger) = \mathcal{V}_\kappa \mathcal{V}_\kappa^\dagger(d\mathcal{V}_\kappa)\mathcal{V}_\kappa^\dagger + \mathcal{V}_\kappa(d\mathcal{V}_\kappa^\dagger) + \mathcal{V}_\kappa \mathcal{V}_\kappa^\dagger d \tag{6.54}$$

are equal. Hence, squaring

$$\mathcal{V}_\kappa(d+\mathcal{A}_\kappa)^2\mathcal{V}_\kappa^\dagger = \mathcal{V}_\kappa\,(d\mathcal{A}_\kappa)\mathcal{V}_\kappa^\dagger = \mathcal{P}_\kappa\,(d\mathcal{P}_\kappa)\,(d\mathcal{P}_\kappa) \tag{6.55}$$

on using $d^2 = 0$, eq.(6.54) and $\mathcal{P}_\kappa(d\mathcal{P}_\kappa)\mathcal{P}_\kappa = 0$. Thus

$$\int_{S^2}(d\mathcal{A}_\kappa) = \int_{S^2}\operatorname{Tr}\,\mathcal{V}_\kappa(d\mathcal{A}_\kappa)\mathcal{V}_\kappa^\dagger = \int_{S^2}\operatorname{Tr}\,\mathcal{P}_\kappa\,(d\mathcal{P}_\kappa)\,(d\mathcal{P}_\kappa)\ . \tag{6.56}$$

We can integrate the LHS. For this we write (taking a section of the bundle $U(1) \to S^3 \to S^2$ over $S^2\backslash\{\text{north pole}(0,0,1)\}$),

$$z(x) = e^{-i\tau_3\varphi/2}e^{-i\tau_2\theta/2}e^{-i\tau_3\varphi/2}\begin{pmatrix}1\\0\end{pmatrix} = \begin{pmatrix}e^{-i\varphi}\cos\frac{\theta}{2}\\ \sin\frac{\theta}{2}\end{pmatrix}\ . \tag{6.57}$$

Taking into account the fact that $U(\vec{x})$ is independent of φ at $\theta = 0$, we get

$$\int_{S^2}(d\mathcal{A}_\kappa) = -\int e^{i\kappa\varphi}\,de^{-i\kappa\varphi} = 2\pi i\kappa\ . \tag{6.58}$$

This and eq.(6.56) reproduce eq.(6.6).

The Belavin-Polyakov bound [79] for S_κ can now be got from the inequality

$$\operatorname{Tr}\mathcal{C}_{\kappa,i}^\dagger\mathcal{C}_{\kappa,i} \geq 0\ ,\quad \mathcal{C}_{\kappa,i} = \mathcal{W}_\kappa\mathcal{B}_{\kappa,i}\mathcal{V}_\kappa^\dagger \pm \mathcal{W}_\kappa(\epsilon_{ijl}x_j\mathcal{B}_{\kappa,l})\mathcal{V}_\kappa^\dagger\ . \tag{6.59}$$

6.4.1   *Relation Between* $\mathcal{P}^{(\kappa)}$ *and* $\mathcal{P}_\kappa$

In the treatment in [81], for $\kappa > 0$, the fuzzy σ-model was based on the continuum projector

$$P^{(\kappa)}(x) = P_1(x) \otimes ... \otimes P_1(x) = \prod_{i=1}^{\kappa} \frac{1}{2}(1 + \tau^{(i)} \cdot x) \qquad (6.60)$$

and its unitary transform

$$\mathcal{P}^{(\kappa)}(x) = U^{(\kappa)}(x)P^{(\kappa)}(x)U^{(\kappa)}(x)^{-1} , \; U^{(\kappa)}(x) = U(x)\otimes...U(x) \; (\kappa \text{ factors}).$$
$$(6.61)$$

At each x, the stability group of $P^{(\kappa)}(x)$ is $U(1)$ with generator $\frac{1}{2}\sum_{i=1}^{\kappa}\tau^{(i)} \cdot x$, and we get a sphere S^2 as $U(x)$ is varied. Thus $U^{(\kappa)}(x)$ gives a section of a sphere bundle over a sphere, leading us to identify $\mathcal{P}^{(\kappa)}$ with a $\mathbb{C}P^1$-field. Furthermore, the R.H.S. of eq.(6.56) (with $\mathcal{P}^{(\kappa)}$ replacing $\mathcal{P}_\kappa$) gives κ as the invariant associated with $\mathcal{P}^{(\kappa)}$, suggesting a correspondence between κ and winding number.

We can write

$$\mathcal{P}^{(\kappa)} = \mathcal{V}^{(\kappa)}\mathcal{V}^{(\kappa)\dagger} , \quad \mathcal{V}^{(\kappa)} = \mathcal{V}_1 \otimes ... \otimes \mathcal{V}_1 \quad (\kappa\text{factors}), \qquad (6.62)$$

its connection $\mathcal{A}^{(\kappa)}$ and an action as previously. A computation similar to the one leading to eq.(6.56) shows that

$$-\frac{i}{2\pi}\int d\mathcal{A}^{(\kappa)} = \kappa . \qquad (6.63)$$

So κ is the Chern invariant of the projective module associated with $\mathcal{P}^{(\kappa)}$.

For $\kappa < 0$, we must change x to $-x$ in (6.60), and accordingly change other expressions.

We note that κ cannot be identified with the winding number of the map $x \rightarrow \mathcal{P}^{(\kappa)}(x)$. To see this, say for $\kappa > 0$, we show that there is a winding number κ map from $\mathcal{P}^{(\kappa)}$ to $\mathcal{P}_\kappa(x)$. As that is also the winding number of the map $x \rightarrow \mathcal{P}_\kappa(x)$, the map $x \rightarrow \mathcal{P}^{(\kappa)}(x)$ must have winding number 1.

The map $\mathcal{P}^{(\kappa)} \rightarrow \mathcal{P}_\kappa(x)$ is induced from the map

$$\mathcal{V}^{(\kappa)} \; \rightarrow \; \mathcal{V}_\kappa = \begin{pmatrix} \mathcal{V}^{(\kappa)}_{11...1} \\ \mathcal{V}^{(\kappa)}_{22...2} \end{pmatrix} \qquad (6.64)$$

and their expressions in terms of $\mathcal{V}^{(\kappa)}$ and $\mathcal{V}_\kappa$. In (6.64) all the points $\mathcal{V}^{(\kappa)}(z_1 e^{2\pi ij/\kappa}, z_2 e^{2\pi il/\kappa})$, $j, l \in \{0, 1, ..., \kappa - 1\}$, have the same image, but in the passage to $\mathcal{P}^{(\kappa)}$ and $\mathcal{P}_\kappa$, the overall phase of z is immaterial. However, the projectors for $\mathcal{V}^{(\kappa)}(z_1 e^{2\pi ij/\kappa}, z_2)$ and $\mathcal{V}^{\dagger}_\kappa(z_1, z_2 e^{2\pi ij/\kappa})$ are distinct and map to the same $\mathcal{P}_\kappa$, giving winding number κ.

We have not understood the relation between the models based on $\mathcal{P}^{(\kappa)}$ and $\mathcal{P}_\kappa$.

6.5 Fuzzy $\mathbb{C}P^1$-Models

The advantage of the preceding formulation using $\{z_\alpha\}$ is that the passage to fuzzy models is relatively transparent. Thus let $\xi = (\xi_1, \xi_2) \in \mathbb{C}^2\backslash\{0\}$. We can then identify z and x as

$$z = \frac{\xi}{|\xi|} \ , \quad |\xi| = \sqrt{|\xi_1|^2 + |\xi_2|^2} \quad x_i = z^\dagger \tau_i z \ . \tag{6.65}$$

Quantization of the ξ's and ξ^*'s consists in replacing ξ_α by annihilation operators a_α and ξ_α^* by $a_\alpha^\dagger$. $|\xi|$ is then the square root of the number operator:

$$\hat{N} = \hat{N}_1 + \hat{N}_2 \ , \quad \hat{N}_1 = a_1^\dagger a_1 \ , \quad N_2 = a_2^\dagger a_2 \ ,$$

$$\hat{z}_\alpha^\dagger = \frac{1}{\sqrt{\hat{N}}} a_\alpha^\dagger = a_\alpha^\dagger \frac{1}{\sqrt{\hat{N}+1}} \ , \quad \hat{z}_\alpha = \frac{1}{\sqrt{\hat{N}+1}} a_\alpha = a_\alpha \frac{1}{\sqrt{\hat{N}}} \ ,$$

$$\hat{x}_i = \frac{1}{\sqrt{\hat{N}}} a^\dagger \tau_i a \ . \tag{6.66}$$

(We have used hats on some symbols to distinguish them as fuzzy operators).

We will apply these operators only on the subspace of the Fock space with eigenvalue $n \geq 1$ of $\hat{N}$, where $\frac{1}{\sqrt{\hat{N}}}$ is well-defined. This restriction is natural and reflects the fact that ξ cannot be zero.

6.5.1 *The Fuzzy Projectors for $\kappa > 0$*

On referring to (6.9), we see that if $\kappa > 0$, for the quantized versions $\hat{v}_\kappa$, $\hat{v}_\kappa^\dagger$ of v_κ, v_κ^*, we have

$$\hat{v}_\kappa = \begin{bmatrix} a_1^\kappa \\ a_2^\kappa \end{bmatrix} \frac{1}{\sqrt{\hat{Z}_\kappa}} \ , \quad \hat{v}_\kappa^\dagger = \frac{1}{\sqrt{\hat{Z}_\kappa}} \left[(a_1^\dagger)^\kappa \ (a_2^\dagger)^\kappa \right] \ , \quad \hat{v}_\kappa^\dagger \hat{v}_\kappa = \mathbb{1} \ ,$$

$$\hat{Z}_\kappa = \hat{Z}_\kappa^{(1)} + \hat{Z}_\kappa^{(2)} \ , \quad \hat{Z}_\kappa^{(\alpha)} = \hat{N}_\alpha(\hat{N}_\alpha - 1)...(\hat{N}_\alpha - \kappa + 1) \ . \tag{6.67}$$

The fuzzy analogue of U is a 2×2 unitary matrix $\hat{U}$ whose entries $\hat{U}_{ij}$ are polynomials in $a_a^\dagger a_b$. As for $\hat{\mathcal{V}}_\kappa$, the quantized version of $\mathcal{V}_\kappa$, it is just

$$\hat{\mathcal{V}}_\kappa = \hat{U} \hat{v}_\kappa \tag{6.68}$$

and fulfills

$$\hat{\mathcal{V}}_\kappa^\dagger \hat{\mathcal{V}}_\kappa = \mathbb{1} \ , \tag{6.69}$$

$\hat{\mathcal{V}}_\kappa^\dagger$ being the quantized version of $\mathcal{V}_\kappa^\dagger$. We thus have the fuzzy projectors

$$\hat{P}_\kappa = \hat{v}_\kappa \hat{v}_\kappa^\dagger \ , \quad \hat{\mathcal{P}}_\kappa = \hat{\mathcal{V}}_\kappa \hat{\mathcal{V}}_\kappa^\dagger \ . \tag{6.70}$$

Unlike $\hat{v}_\kappa$, $\hat{\mathcal{V}}_\kappa$ and their adjoints, $\hat{P}_\kappa$ and $\hat{\mathcal{P}}_\kappa$ commute with the number operator $\hat{N}$. So we can formulate a finite-dimensional matrix model for these projectors as follows. Let $\mathcal{F}_n$ be the subspace of the Fock space where $\hat{N} = n$. It is of dimension $n + 1$, and carries the $SU(2)$ representation with angular momentum $n/2$, the $SU(2)$ generators being

$$L_i = \frac{1}{2} a^\dagger \tau_i a \ . \tag{6.71}$$

Its standard orthonormal basis is $|\frac{n}{2}, m >$, $m = -\frac{n}{2}, -\frac{n}{2} + 1, ..., \frac{n}{2}$. Now consider $\mathcal{F}_n \otimes_{\mathbb{C}} \mathbb{C}^2 := \mathcal{F}_n^{(2)}$, with elements $f = (f_1, f_2)$, $f_a \in \mathcal{F}_n$. Then $\hat{P}_\kappa$, $\hat{\mathcal{P}}_\kappa$ act on $\mathcal{F}_n^{(2)}$ in the natural way. For example

$$f \to \hat{\mathcal{P}}_\kappa f, \quad (\hat{\mathcal{P}}_\kappa f)_a = (\hat{\mathcal{P}}_\kappa)_{ab} f_b = (\hat{\mathcal{V}}_{\kappa,a} \hat{\mathcal{V}}_{\kappa,b}^\dagger) f_b \ . \tag{6.72}$$

We can now write explicit matrices for $\hat{P}_\kappa$ and $\hat{\mathcal{P}}_\kappa$. We have:

$$\hat{P}_\kappa = \begin{pmatrix} a_1^\kappa \frac{1}{\hat{Z}_\kappa} a_1^{\dagger\,\kappa} & a_1^\kappa \frac{1}{\hat{Z}_\kappa} a_2^{\dagger\,\kappa} \\ a_2^\kappa \frac{1}{\hat{Z}_\kappa} a_1^{\dagger\,\kappa} & a_2^\kappa \frac{1}{\hat{Z}_\kappa} a_2^{\dagger\,\kappa} \end{pmatrix} \ , \tag{6.73}$$

$$a_1^\kappa \frac{1}{\hat{Z}_\kappa} = \frac{1}{(\hat{N}_1 + \kappa)...(\hat{N}_1 + 1) + \hat{Z}^{(2)}} a_1^\kappa \ , \quad a_1^\kappa a_1^{\dagger\,\kappa} = (\hat{N}_1 + \kappa)...(\hat{N}_1 + 1) \ ,$$

from which its matrix $\hat{P}_\kappa(n)$ for $\hat{N} = n$ can be obtained.

The matrix $\hat{\mathcal{P}}_\kappa(n)$ of $\hat{\mathcal{P}}_\kappa$ is the unitary transform $\hat{U} \hat{P}_\kappa(n) \hat{U}^\dagger$ where $\hat{U}$ is a 2×2 matrix and $\hat{U}_{ab}$ is itself an $(n + 1) \times (n + 1)$ matrix. As for the fuzzy analogue of $\mathcal{L}_i$, we define it by

$$\mathcal{L}_i \hat{\mathcal{P}}_\kappa(n) = [L_i, \hat{\mathcal{P}}_\kappa(n)] \ . \tag{6.74}$$

The fuzzy action

$$S_{F,\kappa}(n) = \frac{c}{2(n+1)} \text{Tr}_{\hat{N}=n} (\mathcal{L}_i \hat{\mathcal{P}}_\kappa(n))^\dagger (\mathcal{L}_i \hat{\mathcal{P}}_\kappa(n)) \quad c = \text{constant} \ , \tag{6.75}$$

follows, the trace being over the space $\mathcal{F}_n^{(2)}$.

6.5.2 *The Fuzzy Projectors for $\kappa < 0$*

For $\kappa < 0$, following an early indication, we must suitably exchange the roles of a_a and $a_a^\dagger$.

6.5.3 The Fuzzy Winding Number

In the literature [80], there are suggestions on how to extend (6.6) to the fuzzy case. They do not lead to an integer value for this number except in the limit $n \to \infty$.

There is also an approach to topological invariants using the Dirac operator and cyclic cohomology. Elsewhere this approach was applied to the fuzzy case [81, 82] and gave integer values, and even a fuzzy analogue of the Belavin-Polyakov bound. However they were not for the action $S_{F,\kappa}$, but for an action which approaches it as $n \to \infty$. In the subsection below, we present an alternative approach to this bound which works for $S_{F,\kappa}$. It looks like (6.28), except that κ becomes an integer only in the limit $n \to \infty$.

There is also a very simple way to associate an integer to $\hat{\mathcal{V}}_\kappa$ [80, 89, 82]. It is equivalent to the Dirac operator approach. We can assume that the domain of $\hat{\mathcal{V}}_\kappa$ are vectors with a fixed value n of $\hat{N}$. Then after applying $\hat{\mathcal{V}}_\kappa$, n becomes $n - \kappa$ if $\kappa > 0$ and $n + |\kappa|$ is $\kappa < 0$. Thus κ is just the difference in the value of $\hat{N}$, or equivalently twice the difference in the value of the angular momentum, between its domain and its range.

We conclude this section by deriving the bound for $S_{F,\kappa}(n)$.

6.5.4 The Generalized Fuzzy Projector : Duality or BPS States

We introduced the projectors $P_\kappa(\cdot, \eta, \lambda)$ and their fields $n^{(\kappa)}(\cdot, \eta, \lambda)$ earlier. [See (6.15), (6.20)] They describe solitons localized at $\frac{x_1 + ix_2}{1 + x_3} = \eta$ and a shape and width controlled by λ. As inspection shows, they are very easy to quantize by replacing ξ_i by a_i and $\bar{\xi}_j$ by $a_j^\dagger$.

The fields $n^{(\kappa)}(\cdot, \eta, \lambda)$ and their projectors $P_\kappa(\cdot, \eta, \lambda)$ have a particular significance. $P_{|\kappa|}(\cdot, \eta, \lambda)$ saturates the bounds (6.30) with the plus sign, $P_{-\kappa}(\cdot, \eta, \lambda)$ saturates it with the minus sign. This result is due to their holomorphicity (anti-holomorphicity) properties as has been explained elsewhere [66].

It is very natural to identify their fuzzy versions as fuzzy BPS states. But as we note below, they do not saturate the bound on the fuzzy action.

6.5.5 The Fuzzy Bound

A proper generalization of the Belavin-Polyakov bound to its fuzzy version involves a slightly more elaborate approach. This is because the straightforward fuzzification of $\vec{\sigma} \cdot \vec{x}$ and $\vec{\tau} \cdot \vec{n}^{(\kappa)}$ and their corresponding projectors do not commute, and the product of such fuzzy projectors is not a projector. *We use this elaborated approach only in this section.* It is not needed elsewhere. In any case, what is there in other sections is trivially adapted to this formalism.

The operators $a_\alpha^\dagger a_\beta$ acting on the vector space with $\hat{N} = n$ generate the algebra $Mat(n+1)$ of $(n+1) \times (n+1)$ matrices. The extra structure comes from regarding them not as observables, but as a Hilbert space of matrices m, m', ... with scalar product $(m', m) = \frac{1}{n+1} \mathrm{Tr}_{\mathbb{C}^{n+1}} m'^\dagger m$, with the observables acting thereon.

To each $\alpha \in Mat(n+1)$, we can associate two linear operators $\alpha^{L,R}$ on $Mat(n+1)$ according to

$$\alpha^L m = \alpha m \ , \quad \alpha^R m = m\alpha \ , \quad m \in Mat(n+1) \ . \tag{6.76}$$

$\alpha^L - \alpha^R$ has a smooth commutative limit for operators of interest. It actually vanishes, and $\alpha^{L,R} \to 0$ if α remains bounded during this limit.

Consider the angular momentum operators $L_i \in Mat(n+1)$. The associated 'left' and 'right' angular momenta $L_i^{L,R}$ fulfil

$$(L_i^L)^2 = (L_i^R)^2 = \frac{n}{2}(\frac{n}{2}+1) \ . \tag{6.77}$$

We now regard a_α, $a_\alpha^\dagger$ of section 6.5.1 as left operators a_α^L and $a_\alpha^{\dagger L}$. $\hat{\mathcal{P}}_\kappa^L$ thus becomes a 2×2 matrix with each entry being a left multiplication operator. It is the linear operator $\hat{\mathcal{P}}_\kappa^L$ on $Mat(n+1) \otimes \mathbb{C}^2$. We tensor this vector space with another $\mathbb{C}^2$ as before to get $\mathcal{H} = Mat(n+1) \otimes \mathbb{C}^2 \otimes \mathbb{C}^2$, with σ_i acting on the last $\mathbb{C}^2$, and $\sigma \cdot \mathcal{L} \hat{\mathcal{P}}_\kappa^L$ denoting the operator $\sigma_i (\mathcal{L}_i \hat{\mathcal{P}}_\kappa)^L$.

We can repeat the previous steps if there are fuzzy analogues γ and Γ of continuum 'world volume' and 'target space' chiralities $\vec{\sigma} \cdot \vec{x}$ and $\vec{\tau} \cdot \vec{n}^{(\kappa)}$ which mutually commute. Then $\frac{1}{2}(1\pm\gamma)$, $\frac{1}{2}(1\pm\Gamma)$ are commuting projectors and the expressions derived at the end of Section 3 generalize, as we shall see.

There is such a γ, due to Watamuras[69], and discussed further by [81]. Following [81], we take

$$\gamma \equiv \gamma^L = \frac{2\sigma \cdot L^L + 1}{n+1} \ . \tag{6.78}$$

The index L has been put to emphasize its left action on $Mat(n+1)$.

As for Γ, we can do the following. $\hat{\mathcal{P}}_\kappa$ acts on the left on $Mat(n+1)$, let us call it $\hat{\mathcal{P}}_\kappa^L$. It has a $\hat{\mathcal{P}}_\kappa^R$ acting on the right and an associated

$$\Gamma \equiv \Gamma_\kappa^R = 2\hat{\mathcal{P}}_\kappa^R - 1 \ , \quad (\Gamma_\kappa^R)^2 = 1 \ . \tag{6.79}$$

As it acts on the right and involves τ's while γ acts on the left and involves σ's,

$$\gamma^L \Gamma^R_\kappa = \Gamma^R_\kappa \gamma^L . \tag{6.80}$$

The bound for (6.75) now follows from

$$\mathrm{Tr}_{\mathcal{H}} \left(\frac{1+\epsilon_1 \gamma^L}{2} \frac{1+\epsilon_2 \Gamma^R_\kappa}{2} \sigma \cdot \mathcal{L}\hat{\mathcal{P}}^L_\kappa \right)^{\dagger} \left(\frac{1+\epsilon_1 \gamma^L}{2} \frac{1+\epsilon_2 \Gamma^R_\kappa}{2} \sigma \cdot \mathcal{L}\hat{\mathcal{P}}^L_\kappa \right) \geq 0 \tag{6.81}$$

$(\epsilon_1, \epsilon_2 = \pm 1)$, and reads

$$\begin{aligned}
S_{F,\kappa} &= \frac{c}{4(n+1)} \mathrm{Tr}_{\mathcal{H}} (\sigma \cdot \mathcal{L}\hat{\mathcal{P}}^L_\kappa)^{\dagger} (\sigma \cdot \mathcal{L}\hat{\mathcal{P}}^L_\kappa) \\
&\geq \frac{c}{4(n+1)} \mathrm{Tr}_{\mathcal{H}} \left((\epsilon_1 \gamma^L + \epsilon_2 \Gamma^R_\kappa)(\sigma \cdot \mathcal{L}\hat{\mathcal{P}}^L_\kappa)(\sigma \cdot \mathcal{L}\hat{\mathcal{P}}^L_\kappa) \right) \\
&\quad + \frac{c}{4(n+1)} \mathrm{Tr}_{\mathcal{H}} \left(\epsilon_1 \epsilon_2 \gamma^L \Gamma^R (\sigma \cdot \mathcal{L}\hat{\mathcal{P}}^L_\kappa)(\sigma \cdot \mathcal{L}\hat{\mathcal{P}}^L_\kappa) \right) .
\end{aligned} \tag{6.82}$$

The analogue of the first term on the R.H.S. is zero in the continuum, being absent in (6.28), but not so now. As $n \to \infty$, (6.82) reproduces (6.28) to leading order n, but has corrections which vanish in the large n limit.

A minor clarification: if τ's are substituted by σ's in $2\hat{\mathcal{P}}^L_1 - 1$, then it is γ^L. The different projectors are thus being constructed using the same principles.

6.6 $\mathbb{C}P^N$-Models

We need a generalization of the Bott projectors to adapt the previous approach to all $\mathbb{C}P^N$.

Fortunately this can be easily done. The space $\mathbb{C}P^N$ is the space of $(N+1) \times (N+1)$ *rank 1* projectors. The important point is the rank. So we can write

$$\mathbb{C}P^N = \langle U^{(N+1)} P_0 U^{(N+1)\dagger} : P_0 = \mathrm{diag.} \underbrace{(0, ..., 0, 1)}_{N+1 \ entries} U^{(N+1)} \in U(N+1) \rangle . \tag{6.83}$$

As before, let $z = (z_1, z_2)$, $|z_1|^2 + |z_2|^2 = 1$, and $x_i = z^{\dagger} \tau_i z$. Then we define

$$v^{(N)}_\kappa(z) = \begin{pmatrix} z^\kappa_1 \\ z^\kappa_2 \\ 0 \\ \cdot \\ \cdot \\ 0 \end{pmatrix} \frac{1}{\sqrt{Z_\kappa}} , \ \kappa > 0; \quad v^{(N)}_\kappa(z) = \begin{pmatrix} z^{*\kappa}_1 \\ z^{*\kappa}_2 \\ 0 \\ \cdot \\ \cdot \\ 0 \end{pmatrix} \frac{1}{\sqrt{Z_\kappa}} , \ \kappa < 0 . \tag{6.84}$$

Since

$$v_\kappa^{(N)}(z)^\dagger v_\kappa^{(N)}(z) = 1 \,,$$

$$P_\kappa^{(N)}(x) = v_\kappa^{(N)}(z) v_\kappa^{(N)}(z)^\dagger \in \mathbb{C}P^N \,. \qquad (6.85)$$

We can now easily generalize the previous discussion, using $P_\kappa^{(N)}$ for P_κ and $U^{(N+1)}$ for U, and subsequently quantizing z_α, z_α^*. In that way we get fuzzy $\mathbb{C}P^N$-models.

$\mathbb{C}P^N$-models can be generalized by replacing the target space by a general Grassmannian or a flag manifold. They can also be elegantly formulated as gauge theories [72]. But we are able to formulate only a limited class of such manifolds in such a way that they can be made fuzzy. The natural idea would be to look for several vectors

$$v_{k_i}^{(N)(i)}(z) \,, \quad i = 1, .., N \qquad (6.86)$$

in $(N + 1)$-dimensions which are normalized and orthogonal,

$$v_{k_i}^{(N)(i)\dagger}(z) v_{k_j}^{(N)(j)}(z) = \delta_{ij} \,, \qquad (6.87)$$

and have the equivariance property

$$v_{k_i}^{(N)(i)}(ze^{i\theta}) = v_{k_i}^{(N)(i)}(z) e^{i\,k_i\theta} \,. \qquad (6.88)$$

The orbit of the projector $\sum_{i=1}^M v_{k_i}^{(N)(i)}(z) v_{k_i}^{(N)(i)\dagger}(z)$ under $U^{(N+1)}$ will then be a Grassmannian for each $M \leq N$, while the orbit of $\sum_i \lambda_i v_{k_i}^{(N)(i)}(z) v_{k_i}^{(N)(i)}(z)^\dagger$ with possibly unequal λ_i under $U^{(N+1)}$ will be a flag manifold.

But we can find such $v_{k_i}^{(N)(i)}$ only for $i = 1, 2, ..., M \leq \frac{N+1}{2}$.

For instance in an $(N + 1) = 2L$-dimensional vector space, for integer L, we can form the vectors

$$v_{k_1}^{(N)(1)}(z) = \begin{pmatrix} z_1^{k_1} \\ z_2^{k_1} \\ 0 \\ . \\ 0 \end{pmatrix} \frac{1}{\sqrt{Z_{k_1}}} \,, \quad v_{k_2}^{(N)(2)}(z) = \begin{pmatrix} 0 \\ 0 \\ z_1^{k_2} \\ z_2^{k_2} \\ 0 \\ . \\ 0 \end{pmatrix} \frac{1}{\sqrt{Z_{k_2}}}, \cdots ,$$

$$v_{k_L}^{(N)(L)}(z) = \begin{pmatrix} 0 \\ . \\ 0 \\ z_1^{k_L} \\ z_2^{k_L} \end{pmatrix} \frac{1}{\sqrt{Z_{k_L}}} \qquad (6.89)$$

for $k_i > 0$. For those k_i which are negative, we replace $v_{k_i}^{(N)(i)}(z)$ here by $v_{|k_i|}^{(N)(i)}(z)^*$:

$$v_{k_i}^{(N)(i)}(z) = v_{|k_i|}^{(N)(i)}(z)^* \ , \ k_i < 0 \ . \tag{6.90}$$

These $v_{k_i}^{(N)(i)}$ are orthonormal for all z with $\sum_\alpha |z_\alpha|^2 = 1$, so that we can handle Grassmannians and flag manifolds involving projectors up to rank L.

If N instead is $2L$, we can write

$$v_{k_1}^{(N)(1)}(z) = \begin{pmatrix} z_1^{k_1} \\ z_2^{k_1} \\ 0 \\ \cdots \\ 0 \end{pmatrix} \frac{1}{\sqrt{Z_{k_1}}} \ , \ v_{k_2}^{(N)(2)}(z) = \begin{pmatrix} 0 \\ 0 \\ z_1^{k_2} \\ z_2^{k_2} \\ 0 \\ \cdots \\ 0 \end{pmatrix} \frac{1}{\sqrt{Z_{k_2}}}, \cdots ,$$

$$v_{k_L}^{(N)(L)}(z) = \begin{pmatrix} 0 \\ . \\ 0 \\ z_1^{k_L} \\ z_2^{k_L} \\ 0 \end{pmatrix} \frac{1}{\sqrt{Z_{k_L}}} \tag{6.91}$$

for $k_i > 0$, and use (6.90) for $k_i < 0$.

But we can find no vector $v_{k_{L+1}}^{(N)(L+1)}(z)$ fulfilling

$$v_{k_i}^{(N)(i)}(z)^\dagger v_{k_{L+1}}^{(N)(L+1)}(z) = \delta_{i,L+1}, \quad i = 1, 2, .., L+1 \ ,$$

$$v_{k_{L+1}}^{(N)(L+1)}(ze^{i\theta}) = v_{k_{L+1}}^{(N)(L+1)}(z)e^{ik_{L+1}\theta} \ . \tag{6.92}$$

The quantization or fuzzification of these models can be done as before. But lacking suitable $v_{k_i}^{(i)}$ for $i > L$, the method fails if the target flag manifold involves projectors of rank $> \frac{N+1}{2}$.

Note that we cannot consider vectors like

$$v'(z) = \begin{pmatrix} 0 \\ . \\ 0 \\ z_i^k \\ 0 \\ . \\ 0 \end{pmatrix} \frac{1}{|z_i|^k} \ , \quad k > 0 \ , \ i = 1 \text{ or } 2 \tag{6.93}$$

and $v'(z)^*$. That is because z_i can vanish compatibly with the constraint $|z_1|^2 + |z_2|^2 = 1$, and $v'(z)$, $v'(z)^*$ are ill-defined when $z_i = 0$.

The flag manifolds are coset spaces $\mathcal{M} = SU(K)/S[U(k_1) \otimes U(k_2) \otimes .. \otimes U(k_\sigma)]$, $\sum k_i = K$. Since $\pi_2(\mathcal{M}) = \underbrace{\mathbb{Z} \oplus ... \oplus \mathbb{Z}}_{\sigma - 1 \ terms}$, a soliton on $\mathcal{M}$ is now characterized by $\sigma - 1$ winding numbers, with each number allowed to take either sign. [We get $(\sigma - 1)$ and not σ winding numbers. That is because, $det(u_1 \otimes u_2 \otimes \cdots \otimes u_{(k_\sigma)}) = 1$ if $u_i \in U(k_i)$, and hence $\sum_{i=1}^{\sigma} n_i = 0$ if n_i is the winding number associated with $U(k_i)$.] The two possible signs for k_i in $v_{k_i}^{(i)}$ reflect this freedom.

Chapter 7

Fuzzy Gauge Theories

Gauge transformations on commutative spaces are based on transformations which depend on space-time points P. Thus if G is a conventional global group, the associated gauge group is the group of maps $\mathcal{G}$ from space-time to $\mathcal{G}$, the group multiplication being point-wise multiplication. For each irreducible representation (IRR) σ of G, there is an IRR Σ of $\mathcal{G}$ given by $\Sigma(g \in \mathcal{G})(p) = \sigma(g(p))$. The construction works for any connected Lie group G. There is no problem in composing representations of G either: if Σ_i are representations of $\mathcal{G}$ associated with representations of σ_i of G, then we can define the representations $\Sigma_1 \hat{\otimes} \Sigma_2$ which has the same relation to $\sigma_1 \otimes \sigma_2$ that Σ_i have to σ_i: $\Sigma_1 \hat{\otimes} \Sigma_2(g)(p) = [\sigma_1 \otimes \sigma_2](g(p)) = \sigma_1(g(p)) \otimes \sigma_2(g(p))$. Thus such products of Σ are defined using those of G at each p. Existence of these products is essential to describe gauge theories of particles and fields transforming by different representations of G.

An additional point of significance is that there is no condition on G, except that it is a compact connected Lie group.

For general noncommutative manifolds, several of these essential features of $\mathcal{G}$ are absent. Thus in particular

- Noncommutative manifolds require G to be a $U(N)$ group,
- Only a very limited and quite inadequate number of representations of the gauge group can be defined.

We shall illustrate these points below for the fuzzy gauge groups $\mathcal{G}_F$ based on S_F^2, but one can see the generalities of the considerations.

There is an important map, the Seiberg-Witten(SW), map for a noncommutative deformation of $\mathbb{R}^N$. In that case the deformed algebra $\mathbb{R}_\theta^N$ depends continuously on a parameter θ, becoming the commutative algebra for $\theta = 0$. If a certain gauge group on $\mathbb{R}_\theta^N$ is $\mathcal{G}_\theta$, it becomes a standard gauge group $\mathcal{G}_0$ on $\mathbb{R}_0^N = \mathbb{R}^N$. The SW map is based on a homomorphism from $\mathbb{R}_\theta^N$ to $\mathbb{R}_0^N$ and connects gauge theories for different θ. The aforementioned problems can be more or less overcome on $\mathbb{R}_\theta^N$ using this map.

But fuzzy spheres have no continuous parameter like θ. What plays the role of θ is $\frac{1}{L}$ where $2L$ is the cut-off angular momentum, and $\frac{1}{L}$ assumes discrete values. Fuzzy spheres have no SW map as originally conceived, and we can not circumvent its gauge-theoretic problems along the lines for $\mathbb{R}_\theta^N$.

There is however a complementary positive feature of fuzzy spaces. While S_F^2 for example presents problems in describing particles of charge $\frac{1}{3}$ and $\frac{2}{3}$ at the same time (because we cannot "tensor" representations of the fuzzy $U(1)$ gauge group $\mathcal{G}_F(U(1))$), we can describe particles with differing magnetic charges. The projective modules for all magnetic charges were already explained in Chapter 5 and 6. There is no symmetry ("duality") here between electric and magnetic charges.

7.1 Limits on Gauge Groups

The conditions on gauge groups on the fuzzy sphere arise algebraically. They can be understood at the Lie algebraic level.

If $\{\lambda_a\}$ is a basis for the Lie algebra of G in a representation σ, the Lie algebra of $\mathcal{G}_F$, the fuzzy gauge group of G, are generated by

$$\lambda_a \xi_a \qquad (7.1)$$

where ξ_a are $(2L+1) \times (2L+1)$ matrices. ξ_a become functions on S^2 in the large L-limit.

Now consider the commutator

$$[\lambda_a \xi_a , \lambda_b \eta_b], \qquad \eta_b = (2L+1) \times (2L+1) \text{ matrix} \qquad (7.2)$$

of two such Lie algebra elements. We get

$$[\lambda_a , \lambda_b]\xi_a \eta_b + \lambda_a \lambda_b[\xi_a , \eta_b] = iC_{ab}^c \xi_a \eta_b \lambda_c + \lambda_a \lambda_b[\xi_a , \eta_b],$$
$$C_{ab}^c = \text{structure constants of the Lie algebra of } \mathcal{G}. \qquad (7.3)$$

Since $C_{ab}^c \xi_a \eta_b \in S_F^2$, the first term is of the appropriate form for a fuzzy gauge group of G. But the last term is not, it involves $\lambda_a \lambda_b$ which is a product of two generators of G. By taking repeated commutators, we will generate products of all orders of λ_a's and their commutators. If σ is irreducible and of dimension d, we will get all the $d \times d$ hermitian matrices in this way and not just the λ_a. This means that the fuzzy gauge group is that of $U(d)$.

In the commutative limit, $[\xi_a, \eta_b]$ is zero and this problem does not occur.

This escalation of the gauge group to $U(d)$ is difficult to control. No convincing proposal to minimize its effects exists. [See in this connection [84–86]].

In any case, $U(d)$ gauge theories without matter fields can be consistently formulated on fuzzy spheres.

For applications, there is one mitigating circumstance: In the standard model, if we gauge just $SU(3)_C$ and $U(1)_{EM}$, namely the $SU(3)$ of colour and $U(1)$ of electromagnetism, the group is actually $U(3)$ [88]. Likewise, the weak group is not $SU(2) \times U(1)$, but $U(2)$. Thus gauge fields without matter in these sectors can be studied on fuzzy spheres.

Unfortunately, this does not mean that these gauge theories can be formulated satisfactorily on S_F^2 or (for a four-dimensional continuum limit) on $S_F^2 \times S_F^2$ say, when quarks and leptons are included. For example with different flavours, different charges like $2/3$ and $-1/3$ occur, and there is no good way to treat arbitrary representations of gauge groups in noncommutative geometry [86]. We explain this problem now.

7.2 Limits on Representations of Gauge Groups

For the fuzzy $U(d)$ gauge group on the fuzzy sphere $S_F^2(2L+1)$, we consider $S_F^2(2L+1) \otimes \mathbb{C}^d$. The fuzzy $U(d)$ gauge group $U(d)_F$ consists of $d \times d$ matrices U with coefficients in $S_F^2(2L+1) : U_{ij} \in S_F^2(2L+1)$. The $U(d)_F$ can act in three different ways on $S_F^2(2L+1) \otimes \mathbb{C}^d$: on left, right and both:

   i. Left action : $U \to U^L$ where $U^L X = UX$ for $X \in S_F^2(2L+1) \otimes \mathbb{C}^d$,
   ii. Right action : $U \to (U^\dagger)^R : (U^\dagger)^R X = XU^\dagger$,
   iii. Adjoint action : $U \to AdU : AdU\, X = UXU^\dagger$.

If *i.* gives representation Λ, then *ii.* is its complex conjugate Λ^* and *iii.* is its adjoint representation $Ad\,\Lambda$. We are guaranteed that these representations can always be constructed.

But can we construct other representations such as the one corresponding to $\Sigma_1 \widehat{\otimes} \Sigma_2$? The answer appears to be no.

The reason is as follows $\widehat{\otimes}$ is not the tensor product $\otimes$. In $\Sigma \otimes \Sigma$, we get functions of two variables p and q: $(\Sigma(g) \otimes \Sigma(g))(p,q) = \sigma(g(p)) \otimes \sigma(g(q))$.

We must restrict $(\Sigma(g) \otimes \Sigma(g))$ to the diagonal points (p, p) to get $\widehat{\otimes}$.

In noncommutative geometry, the tensor product $\Lambda_1 \otimes \Lambda_2$ exists of course since $\Lambda_1(U) \otimes \Lambda_2(U)$ is defined, and gives a representation of $U(d)_F$. But noncommutative geometry has no sharp points. That obstructs the construction of an analogue of diagonal points, or the restriction of $\otimes$ to an analogue of $\widehat{\otimes}$.

There exist proposals [86] to get around this problem using Higgs fields. There is also the work of Kürkçüoğlu and Sämann [87] on the use of "twisted" coproducts for defining the action of diffeomorphisms and gauge groups on the fuzzy sphere. It has the potentiality to overcome this problem.

7.3　Connection and Curvature

As a convention we choose the gauge potential to act on the left of $S_F^2(2L + 1) \otimes \mathbb{C}^d$. So the components of the gauge potentials are

$$A_i^L = (A_i^L)^a \lambda_a, \quad (A_i^L)^a \in S_F^2(2l + 1) \tag{7.4}$$

where $\lambda_a, (a = 1, \cdots, d^2)$ are the $d \times d$ basis matrices for the Lie algebra of $U(d)$. They can be the Gell-Mann matrices.

The covariant derivative ∇ is then the usual one:

$$\nabla_i = \mathcal{L}_i + A_i^L \tag{7.5}$$

The curvature is

$$
\begin{aligned}
F_{ij} &= [\nabla_i, \nabla_j] - i\varepsilon_{ijk}\nabla_k \\
&= [\mathcal{L}_i, \mathcal{L}_j] + \mathcal{L}_i A_j^L - \mathcal{L}_j A_i^L + [A_i^L, A_j^L] - i\varepsilon_{ijk}(\mathcal{L}_k + A_k^L) \\
&= \mathcal{L}_i A_j^L - \mathcal{L}_j A_i^L + [A_i^L, A_j^L] - i\varepsilon_{ijk}A_k^L.
\end{aligned}
\tag{7.6}
$$

The subtraction of $i\varepsilon_{ijk}\nabla_k$ is needed to cancel the $[\mathcal{L}_i, \mathcal{L}_j]$ term in $[\nabla_i, \nabla_j]$.

There is one important condition on ∇_i. On S^2, A^L becomes a commutative gauge field a and its components a_i have to be tangent to S^2:

$$x_i a_i = 0. \tag{7.7}$$

We need a condition on ∇_i which becomes this condition for large L.

A simple condition of such a nature is due to Nair and Polychronakos [90] and reads

$$(L_i^L + A_i^L)^2 = L(L + 1). \tag{7.8}$$

This is compatible with gauge invariance. Its expansion is

$$L_i^L A_i^L + A_i^L L_i^L + A_i^L A_i^L = 0\,.$$ (7.9)

We have that $\frac{A_i^L}{L} \to 0$ as $L \to \infty$. Dividing (7.9) by L and passing to the limit, we thus get (7.7).

The fuzzy Yang-Mills action is

$$\mathcal{S}_F = \frac{1}{4e^2} Tr F_{ij}^2 + \lambda(\nabla_i^2 - L(L+1))\,, \quad \lambda \geq 0\,,$$ (7.10)

where the second term is a Lagrange multiplier: it enforces the constraint (7.8) as $\lambda \to \infty$.

There is an approach to fuzzy gauge theories which has connections to string-theoretical matrix models such as those of Ishibashi, Kawai, Kitazawa and Tsuchiya [91] and Alekseev, Recknagel and Schomerus [136]. This approach developed by Watumaras [92], and Steinacker [93] is conceptually important and can also have practical advantages. We refer to the literature for details.

7.4 Instanton Sectors

The above action is good in the sector with no instantons. But $U(d)$ gauge theories on S^2 have instantons, or equivalently, twisted $U(1)$-bundles on S^2. We outline how to incorporate instantons on the fuzzy sphere, taking $d = 1$ for simplicity.

The projective modules for instanton sectors were constructed previously. We review it briefly constructing the modules in a different (but Morita equivalent [36]) manner.

The instanton sectors on S^2 correspond to $U(1)$ bundles thereon. To build the corresponding projective module for Chern number $2T \in \mathbb{Z}^+$, introduce $\mathbb{C}^{2T+1}$ carrying the angular momentum T representation of $SU(2)$. Let T_i be the angular momentum operators in this representation with standard commutation relations. Let $Mat(2L + 1) \otimes \mathbb{C}^{(2T+1)} \equiv Mat(2L + 1)^{(2T+1)}$. We let P^{L+T} be the projector coupling left angular momentum operators L^L and T to produce maximum angular momentum $L + T$. Then the projective module $P^{L+T} Mat(2L + 1)^{(2T+1)}$ is a fuzzy analogue of sections of $U(1)$ bundles on S^2 with Chern number $2T > 0$ [81]. If instead we couple L^L and T to produce the least angular momentum $L - T$ using the projector P^{L-T}, then the projective module $P^{L-T} Mat(2L + 1)^{(2T+1)}$ corresponds to Chern number $-2T$. (We assume that $L \geq T$).

The derivation $\mathcal{L}_i$ does not commute with $P^{L\pm T}$ and has no action on these modules. But, $\mathcal{J}_i = \mathcal{L}_i + T_i$ does commute with $P^{L\pm T}$. Thus $\mathcal{L}_i$ must be replaced with $\mathcal{J}_i$ in further considerations. $\mathcal{J}_i$ is to be considered the total angular momentum . The addition of T_i to $\mathcal{L}_i$ here is the algebraic analogue of "mixing of spin and isospin". [71, 70]. It is interesting that the mixing of "spin and isospin" occurs already in our finite-dimensional matrix model and does not need noncompact spatial slices and spontaneous symmetry breaking.

We must next gauge $\mathcal{J}_i$. In the zero instanton sector, the fuzzy gauge fields A_i^L were functions of L_i^L. But that is not possible now since A_i^L does not commute with $P^{L\pm T}$. Instead we require A_i^L to be a function of $\vec{L}^L + \vec{T}$ and write for the covariant derivative

$$\nabla_i = \mathcal{J}_i + A_i^L . \tag{7.11}$$

When $L \to \infty$, $\vec{T}$ can be ignored, and then A_i^L becomes a function of x, just as we want.

The transversality condition must be modified. It is now

$$(L_i^L + T_i + A_i^L)^2 = (L_i^L + T_i)^2 \tag{7.12}$$

where

$$(L_i^L + T_i)^2 = (L \pm T)(L \pm T + 1) \tag{7.13}$$

on $P^{L\pm T} Mat(2L+1)^{(2T+1)}$.

The curvature F_{ij} and the action $\mathcal{S}_F$ are as in (7.6) and (7.10).

7.5 The Partition Function and the θ-parameter

Existence of instanton bundles on a commutative manifold brings in a new parameter, generally called θ, as in QCD. The partition function Z_θ depends on θ.

Let us denote the action in the instanton number $K \in \mathbb{Z}$ sector by $\mathcal{S}_F^K$. Then

$$Z_\theta = \sum_k \int DA_i^L e^{-\mathcal{S}_F^K + iK\theta} . \tag{7.14}$$

We thus have a matrix model for $U(d)$ gluons.

In the continuum, K can be written as the integral of trF (where trace tr (with lower case t) is over the internal indices). In four dimensions it is the integral of $trF \wedge F$. But on S_F^2, $Tr\varepsilon^{ij}F_{ij}$ (where Tr is over both internal indices and S_F^2) is not an integer. A similar difficulty arises for $S_F^2 \times S_F^2$ or $\mathbb{C}P^2$.

In continuum gauge theory, F and $F \wedge F$ play a role in discussions of chiral symmetry breaking. They arise as the local anomaly term in the continuity equation for chiral current. Therefore although Z_θ defines the theory, it is still helpful to have fuzzy analogues of the topological densities TrF and $TrF \wedge F$.

It is possible to construct fuzzy topological densities using cyclic cohomology [34]. We will not review cyclic cohomology here.

Chapter 8

The Dirac Operator and Axial Anomaly

8.1 Introduction

The Dirac operator is central for fundamental physics. It is also central in noncommutative geometry. In Connes' approach [34], it is possible to formulate metrical, differential geometric and bundle-theoretic ideas using the Dirac operator in a form generalisable to noncommutative manifolds.

In this chapter, we explain the theory of the fuzzy Dirac operator basing it on the Ginsparg-Wilson (GW) algebra [94]. This algebra appeared first in the context of lattice gauge theories as a device to write the Dirac operator overcoming the well-known fermion-doubling problem. The same algebra appears naturally for the fuzzy sphere. The theory of the fuzzy Dirac operator can be based on this algebra. It has no fermion doubling and correctly and elegantly reproduces the integrated $U(1)_A$-(axial) anomaly.

Incidentally the association of the GW-algebra with the fuzzy sphere is surprising as the latter is not designed with this algebra in mind.

Below we review the GW-algebra in its generality. We then adapt it to S_F^2. Our discussion here closely follows [95]. Other references on the subject include [105–107].

8.2 A Review of the Ginsparg-Wilson Algebra

In its generality, the Ginsparg-Wilson algebra $\mathcal{A}$ can be defined as the unital $*$-algebra over $\mathbb{C}$ generated by two $*$-invariant involutions Γ and Γ':

$$\mathcal{A} = \left\langle \Gamma, \Gamma' : \quad \Gamma^2 = \Gamma'^2 = \mathbb{1}, \quad \Gamma^* = \Gamma, \quad \Gamma'^* = \Gamma' \right\rangle, \qquad (8.1)$$

($*$ generally denoting the adjoint in any representation). The unity of $\mathcal{A}$ has been indicated by $\mathbb{1}$.

In any such algebra, we can define a Dirac operator

$$D' = \frac{1}{a}\Gamma(\Gamma + \Gamma') \, , \tag{8.2}$$

where a is the "lattice spacing". It fulfills

$$D'^* = \Gamma D' \Gamma, \quad \{\Gamma, D'\} = a D' \Gamma D' \, . \tag{8.3}$$

(8.2) and (8.3) give the original formulation [94]. But they are equivalent to (8.1), since (8.2) and (8.3) imply that

$$\Gamma' = \Gamma(aD') - \Gamma \tag{8.4}$$

is a $*$-invariant involution [98] [96].

Each representation of (8.1) is a particular realization of the Ginsparg-Wilson algebra. Representations of physical interest are reducible.

Here we choose

$$D = \frac{1}{a}(\Gamma + \Gamma') \, , \tag{8.5}$$

instead of D' as our Dirac operator, as it is self-adjoint and has the desired continuum limit.

From Γ and Γ', we can construct the following elements of $\mathcal{A}$:

$$\Gamma_0 = \frac{1}{2}\{\Gamma, \Gamma'\} \, , \tag{8.6}$$

$$\Gamma_1 = \frac{1}{2}(\Gamma + \Gamma') \, , \tag{8.7}$$

$$\Gamma_2 = \frac{1}{2}(\Gamma - \Gamma') \, , \tag{8.8}$$

$$\Gamma_3 = \frac{1}{2i}[\Gamma, \Gamma'] \, . \tag{8.9}$$

Let us first look at the centre $\mathcal{C}(\mathcal{A})$ of $\mathcal{A}$ in terms of these operators. It is generated by Γ_0 which commutes with Γ and Γ' and hence with every element of $\mathcal{A}$. Γ_i^2, $i = 1, 2, 3$ also commute with every element of $\mathcal{A}$, but they are not independent of Γ_0. Rather,

$$\Gamma_1^2 = \frac{1}{2}(\mathbb{1} + \Gamma_0) \, , \tag{8.10}$$

$$\Gamma_2^2 = \frac{1}{2}(\mathbb{1} - \Gamma_0) \, , \tag{8.11}$$

$$\rightarrow \quad \Gamma_1^2 + \Gamma_2^2 = \mathbb{1} \, , \tag{8.12}$$

$$\Gamma_0^2 + \Gamma_3^2 = \mathbb{1} \, . \tag{8.13}$$

Notice also that

$$\{\Gamma_i, \Gamma_j\} = 0 \, , \ i, j = 1, 2, 3, \ i \neq j \, . \tag{8.14}$$

From now on by $\mathcal{A}$ we will mean a representation of $\mathcal{A}$.

The relations (8.10)-(8.13) contain spectral information. From (8.13) we see that

$$-1 \leq \Gamma_0 \leq 1 , \qquad (8.15)$$

where the inequalities mean that the eigenvalues of Γ_0 are accordingly bounded. By (8.10), this implies that the eigenvalues of Γ_1 are similarly bounded.

We now discuss three cases associated with (8.15).

Case 1 :

$\Gamma_0 = 1\!\!1$. Call the subspace where $\Gamma_0 = 1\!\!1$ as V_{+1}. On V_{+1}, $\Gamma_1^2 = 1\!\!1$ and $\Gamma_2 = \Gamma_3 = 0$ by (8.10-8.13). This is subspace of the top modes of the operator $|D|$.

Case 2 :

$\Gamma_0 = -1\!\!1$. Call the subspace where $\Gamma_0 = -1\!\!1$ as V_{-1}. On V_{-1}, $\Gamma_2^2 = 1\!\!1$ and $\Gamma_1 = \Gamma_3 = 0$ by (8.10-8.13). This is the subspace of zero modes of the Dirac operator D.

Case 3 :

$\Gamma_0^2 \neq 1\!\!1$. Call the subspace where $\Gamma_0^2 \neq 1\!\!1$ as V. On this subspace, $\Gamma_i^2 \neq 0$ for $i = 1, 2, 3$ by (8.9-8.12), and therefore

$$sign\,\Gamma_i = \frac{\Gamma_i}{|\Gamma_i|} , \quad |\Gamma_i| = \text{positive square root of } \Gamma_i^2 \qquad (8.16)$$

are well defined and by (8.14) generate a Clifford algebra on V:

$$\{sign\,\Gamma_i, sign\,\Gamma_j\} = 2\delta_{ij} . \qquad (8.17)$$

Consider Γ_2. It anticommutes with Γ_1 and D. Also

$$\text{Tr}\,\Gamma_2 = (\text{Tr}_V + \text{Tr}_{V_{+1}} + \text{Tr}_{V_{-1}})\Gamma_2 , \qquad (8.18)$$

where the subscripts refer to the subspaces over which the trace is taken. These traces can be calculated:

$$\begin{aligned} \text{Tr}_V\,\Gamma_2 &= \text{Tr}_V(sign\,\Gamma_i)\Gamma_2(sign\,\Gamma_i) \quad (i \text{ fixed}, \neq 2) \\ &= -\,\text{Tr}_V\,\Gamma_2 \quad \text{by(8.17)} \\ &= 0, \qquad (8.19) \end{aligned}$$

$$\text{Tr}_{V_{+1}}\,\Gamma_2 = 0, \quad \text{as } \Gamma_2 = 0 \text{ on } V_{+1} . \qquad (8.20)$$

So

$$\text{Tr}\,\Gamma_2 = \text{Tr}_{V_{-1}}\Gamma_2 = \text{Tr}_{V_{-1}}\left(\frac{1+\Gamma_2}{2} - \frac{1-\Gamma_2}{2}\right) = \text{ index of }\Gamma_1\ . \qquad (8.21)$$

Following Fujikawa [96], we can use Γ_2 as the generator of chiral transformations. It is not involutive on $V \oplus V_{+1}$

$$\Gamma_2^2 = \mathbb{1} - \frac{\mathbb{1} + \Gamma_0}{2}\ . \qquad (8.22)$$

But this is not a problem for fuzzy physics. In the fuzzy model below, in the continuum limit, $\Gamma_0 \to -\mathbb{1}$ on all states with $|D| \leq$ a fixed 'energy' E_0 independent of a (and is $-\mathbb{1}$ on V_{-1} where $D = 0$). We can see this as follows. $\Gamma_1 = aD$, so that if $|D| \leq E_0$, $\Gamma_1 \to 0$ as $a \to 0$. Hence by (8.10,8.12), $\Gamma_0 \to -\mathbb{1}$ and $\Gamma_2^2 \to \mathbb{1}$ on these levels.

There are of course states, such as those of V_{+1}, on which Γ_2^2 does not go to $\mathbb{1}$ as $a \to 0$. But their (Euclidean) energy diverges and their contribution to functional integrals vanishes in the continuum limit.

We can interpret (8.22) as follows. The chiral charge of levels with $D \neq 0$ gets renormalized in fuzzy physics. For levels with $|D| \leq E_0$, this renormalization vanishes in the naive continuum limit.

We note that the last feature is positive: it resolves a problem faced in [99] for fuzzy spheres, where all the top modes of the Dirac operator had to be projected out because of the insistence that chirality squares to $\mathbb{1}$ on V_{+1} (see below).

For Dirac operators of maximum symmetry on S_F^2, Γ_0 is a function of the conserved total angular momentum $\vec{J}$ as we shall show. It increases with $\vec{J}^2$ so that V_{+1} consists of states of maximum $\vec{J}^2$. As the calculations below show, on $S_F^2 = Mat(2\ell + 1)$, where $a = O\left(\frac{1}{L}\right)$, this maximum value diverges as $a \to 0$.

8.3   Fuzzy Models

8.3.1   *Review of the Basic Fuzzy Sphere Algebra*

Let us briefly recollect the basic algebraic details.

The algebra for the fuzzy sphere characterized by cut-off $2L$ is the full matrix algebra $Mat(2L + 1) \equiv M_{2L+1}$ of $(2L + 1) \times (2L + 1)$ matrices. On M_{2L+1}, the $SU(2)$ Lie algebra acts either on the left or on the right. Call the operators for left action as L_i^L and for right action as L_i^R. We have

$$L_i^L a = L_i a\ , \quad L_i^R a = a L_i\ , \quad a \in Mat(2L + 1)\ , \qquad (8.23)$$

$$[L_i^L, L_j^L] = i\epsilon_{ijk}L_k^L , \quad [L_i^R, L_j^R] = -i\epsilon_{ijk}L_k^R , \quad (L_i^L)^2 = (L_i^R)^2 = L(L+1)\mathbb{1} , \tag{8.24}$$

where L_i is the standard matrix for the i-th component of the angular momentum in the the $(2L + 1)$-dimensional irreducible representation (IRR). The orbital angular momentum which becomes $-i(\vec{r} \wedge \vec{\nabla})_i$ as $L \to \infty$ is

$$\mathcal{L}_i = L_i^L - L_i^R , \quad \mathcal{L}_i a = [L_i, a] . \tag{8.25}$$

As $L \to \infty$, both $\vec{L}^L/L$ and $\vec{L}^R/L$ approach the unit vector $\hat{x}$ with commuting components:

$$\frac{\vec{L}^{L,R}}{L} \xrightarrow{L \to \infty} \hat{x} , \quad \hat{x} \cdot \hat{x} = 1 , \quad [\hat{x}_i, \hat{x}_j] = 0 . \tag{8.26}$$

$\hat{x}$ labels a point on the sphere S^2 in the continuum limit.

8.3.2 The Fuzzy Dirac Operator (No Instantons or Gauge Fields)

Consider $M_{2L+1} \otimes \mathbb{C}^2$. $\mathbb{C}^2$ is the carrier of the spin $1/2$ representation of $SU(2)$ with generators $\frac{1}{2}\sigma_i$, $\sigma_i =$ Pauli matrices. We can couple its spin $1/2$ and the angular momentum L of L_i^L to the value $L + 1/2$. If $(1 + \Gamma)/2$ is the corresponding projector, then [99] [69] [81]

$$\Gamma = \frac{\vec{\sigma} \cdot \vec{L}^L + 1/2}{L + 1/2} . \tag{8.27}$$

Γ is a self-adjoint involution,

$$\Gamma^* = \Gamma \quad , \quad \Gamma^2 = \mathbb{1} . \tag{8.28}$$

There is likewise the projector $(\mathbb{1} + \Gamma')/2$ coupling the spin $1/2$ of $\mathbb{C}^2$ and the right angular momentum $-L_i^R$ to $L + 1/2$, where

$$\Gamma' = \frac{-\vec{\sigma} \cdot \vec{L}^R + 1/2}{L + 1/2} = \Gamma'^* \quad \Gamma'^2 = \mathbb{1} . \tag{8.29}$$

The algebra $\mathcal{A}$ is generated by Γ and Γ'.

The fuzzy Dirac operator of Grosse et al.[7] is

$$D = \frac{1}{a}(\Gamma + \Gamma') = \frac{2}{a}\Gamma_1 = \vec{\sigma} \cdot (\vec{L}^L - \vec{L}^R) + 1 , \quad a = \frac{1}{L + 1/2} . \tag{8.30}$$

Thus the Dirac operator is in this case an element of the Ginsparg-Wilson algebra $\mathcal{A}$.

We can calculate Γ_0 in terms of $\vec{J} = \vec{\mathcal{L}} + \vec{\sigma}/2$:

$$\Gamma_0 = \frac{a^2}{2}[\vec{J}^2 - 2L(L + 1) - \frac{1}{4}] . \tag{8.31}$$

Thus the eigenvalues of Γ_0 increase monotonically with the eigenvalues $j(j+1)$ of $\vec{J}^2$ starting with a minimum for $j = 1/2$ and attaining a maximum of 1 for $j = 2L + 1/2$.

Γ_2 is chirality. It anticommutes with D. For fixed j, as $L \to \infty$, $\Gamma_0 \to -\mathbb{1}$ and $\Gamma_2^2 = \mathbb{1}$ as expected. In fact, Γ_2 in the naive continuum limit is the standard chirality for fixed j. As $L \to \infty$, $\Gamma_2 \to \sigma \cdot \hat{x}$. As mentioned earlier, use of Γ_2 as chirality resolves a difficulty addressed elsewhere [99], where $sign\,(\Gamma_2)$ was used as chirality. That necessitates projecting out V_{+1} and creates a very inelegant situation.

Finally we note that there is a simple reconstruction of Γ and Γ' from their continuum limits [104]. If $\vec{x}$ is not normalized, $\vec{\sigma} \cdot \hat{x} = \frac{\vec{\sigma} \cdot \vec{x}}{|\vec{\sigma} \cdot \vec{x}|}$, $|\vec{\sigma} \cdot \vec{x}| \equiv |((\vec{\sigma} \cdot \vec{x})^2)^{1/2}|$. As $\vec{x}$ can be represented by $\vec{L}^L$ or $\vec{L}^R$ in fuzzy physics, natural choices for Γ and Γ' are $sign\,(\vec{\sigma} \cdot L^L)$ and $-sign\,(\vec{\sigma} \cdot L^R)$. The first operator is $+1$ on vectors having $\vec{\sigma} \cdot \vec{L}^L > 0$ and -1 if instead $\vec{\sigma} \cdot \vec{L}^L < 0$. But if $(\vec{L}^L + \vec{\sigma}/2)^2 = (L + 1/2)(L + 3/2)$, then $\vec{\sigma} \cdot \vec{L}^L = L > 0$, while if $(\vec{L}^L + \vec{\sigma}/2)^2 = (L - 1/2)(L + 1/2)$, $\vec{\sigma} \cdot \vec{L}^L = -(L + 1) < 0$. Γ is $+1$ on former states and -1 on latter states. Thus

$$sign\,(\vec{\sigma} \cdot \vec{L}^L) = \Gamma \ , \tag{8.32}$$

and similarly

$$sign\,(\vec{\sigma} \cdot \vec{L}^R) = -\Gamma' \ . \tag{8.33}$$

It is easy to calculate the spectrum of D. We can write

$$aD = \vec{J}^2 - \vec{\mathcal{L}}^2 - \frac{3}{4} + 1 \tag{8.34}$$

We observe that $[\vec{J}^2 , \vec{\mathcal{L}}^2] = 0$. The spectrum of $\vec{\mathcal{L}}^2$ is

$$spec\,\vec{\mathcal{L}}^2 = \{\ell(\ell + 1) : \ell = 0, 1, \cdots, 2L\} \ , \tag{8.35}$$

whereas that of $\vec{J}^2$ is

$$spec\,\vec{J}^2 = \left\{ j(j + 1) : j = \frac{1}{2}, \frac{3}{2}, \cdots, 2L + \frac{1}{2} \right\} \ . \tag{8.36}$$

Here each j can come from $\ell = j \pm \frac{1}{2}$ by adding spin, except $j = 2L + \frac{1}{2}$ which comes only from $\ell = 2L$. It follows that the eigenvalue of D for $\ell = j - \frac{1}{2}$ is $j + \frac{1}{2} = \ell + 1, \ell \leq 2L$ and for $\ell = j + \frac{1}{2}$ is $-(j + \frac{1}{2}) = -\ell, \ell \leq 2L$.

The spectrum found here agrees *exactly* with what is found in the continuum for $j \leq 2L - \frac{3}{2}$. For $j = 2L + \frac{1}{2}$ we get the positive eigenvalue correctly, but the negative one is missing. That is an edge effect caused by cutting off the angular momentum at $2L$.

8.3.3 The Fuzzy Gauged Dirac Operator (No Instanton Fields)

We adopt the convention that gauge fields are built from operators on $Mat(2L+1)$ which act by left multiplication. For $U(k)$ gauge theory, we start from $Mat(2L+1) \otimes \mathbb{C}^k$. The fuzzy gauge fields A_i^L are $k \times k$ matrices $[(A_i^L)_{mn}]$ where each entry is the operator of left-multiplication by $(A_i)_{mn} \in Mat(2L+1)$ on $Mat(2L+1)$. A_i^L thus acts on $\xi = (\xi_1, \ldots, \xi_k)$, $\xi_i \in Mat(2L+1)$ according to

$$(A_i^L \xi)_m = (A_i)_{mn} \xi_n . \tag{8.37}$$

The gauge-covariant derivative is then

$$\nabla_i(A^L) = \mathcal{L}_i + A_i^L = L_i^L - L_i^R + A_i^L . \tag{8.38}$$

Note how only the left angular momentum is augmented by a gauge field.

The hermiticity condition on A_i^L is

$$(A_i^L)^* = A_i^L , \tag{8.39}$$

where

$$((A_i^L)^* \xi)_m = (A_i^*)_{mn} \xi_n , \tag{8.40}$$

$(A_i^*)_{nm}$ being Hermitian conjugate of $(A_i)_{nm}$. The corresponding field strength F_{ij} is defined by

$$[(L+A)_i^L, (L+A)_j^L] = i\epsilon_{ijk}(L+A)_k^L + iF_{ij} . \tag{8.41}$$

There is a further point to attend to. We need a gauge-invariant condition which in the continuum limit eliminates the component of A_i normal to S^2. There are different such conditions, the following simple one was disccussed in chapter 7, (cf. 7.8):

$$(L_i^L + A_i^L)^2 = (L_i^L)^2 = L(L+1) . \tag{8.42}$$

The Ginsparg-Wilson system can be introduced as follows. As Γ squares to $\mathbb{1}$, there are no zero modes for Γ and hence for $\vec{\sigma} \cdot \vec{L}^L + 1/2$. By continuity, for generic $\vec{A}^L$, its gauged version $\vec{\sigma} \cdot (\vec{L}^L + \vec{A}^L) + 1/2$ also has no zero modes. Hence we can set

$$\Gamma(A^L) = \frac{\vec{\sigma} \cdot (\vec{L}^L + \vec{A}^L) + 1/2}{|\vec{\sigma} \cdot (\vec{L}^L + \vec{A}^L) + 1/2|} , \quad \Gamma(A^L)^* = \Gamma(A^L) , \quad \Gamma(A^L)^2 = \mathbb{1} . \tag{8.43}$$

It is the gauged involution which reduces to $\Gamma = \Gamma(0)$ for zero $\vec{A}^L$.

As for the second involution $\Gamma'(A^L)$, we can set

$$\Gamma'(A^L) = \Gamma'(0) \equiv \Gamma' . \qquad (8.44)$$

On following (8.6-8.9), these idempotents generate the Ginsparg-Wilson algebra with operators $\Gamma_\lambda(A^L)$, where $\Gamma_\lambda(0) = \Gamma_\lambda$.

The operators $\vec{L}^{L,R}$ do not individually have continuum limits as their squares $L(L+1)$ diverge as $L \to \infty$. In contrast $\vec{\mathcal{L}}$ and $\vec{A}^L$ do have continuum limits. This was remarked earlier on for the latter, while $\vec{\mathcal{L}}$ just becomes orbital angular momentum.

To see more precisely how $D(A^L)$, the Dirac operator for gauge field A^L, ($D(0)$ being the D of (8.30)), and $\Gamma_2(A^L)$, behave in the continuum limit, we note that from (8.41),(8.42),

$$\left(\vec{\sigma} \cdot (\vec{L}^L + \vec{A}^L) + \frac{1}{2}\right)^2 = (L + \frac{1}{2})^2 - \frac{1}{2}\epsilon_{ijk}\sigma_i F_{ij} . \qquad (8.45)$$

Therefore we have the expansions

$$\frac{1}{|\vec{\sigma} \cdot (\vec{L}^L + \vec{A}^L) + \frac{1}{2}|} = \frac{2}{\sqrt{\pi}} \int_0^\infty ds \, e^{-s^2(\vec{\sigma} \cdot (\vec{L}^L + \vec{A}^L) + \frac{1}{2})^2}$$

$$= \frac{1}{L + \frac{1}{2}} + \frac{1}{4(L + \frac{1}{2})^3}\epsilon_{ijk}\sigma_i F_{jk} + \dots, \qquad (8.46)$$

$$D(A^L) = (2L + 1)\Gamma_1(A^L) =$$

$$\vec{\sigma} \cdot (\vec{L}^L - \vec{L}^R + \vec{A}^L) + 1 + \frac{\vec{\sigma} \cdot (\vec{L}^L + \vec{A}^L) + \frac{1}{2}}{4(L + \frac{1}{2})^2}\epsilon_{ijk}\sigma_k F_{ij} + \dots$$

$$\Gamma_2(A^L) =$$

$$\frac{\vec{\sigma} \cdot (\vec{L}^L + \vec{A}^L) + \frac{1}{2}}{2(L + \frac{1}{2})} - \frac{-\vec{\sigma} \cdot \vec{L}^R + \frac{1}{2}}{2(L + \frac{1}{2})} + \frac{\vec{\sigma} \cdot (\vec{L}^L + \vec{A}^L) + \frac{1}{2}}{8(L + \frac{1}{2})^3}\epsilon_{ijk}\sigma_k F_{ij} + \dots .$$

$$(8.47)$$

So in the continuum limit, $D(A^L) \to \vec{\sigma} \cdot (\vec{\mathcal{L}} + \vec{A}) + 1$, and $\Gamma_2(A) \to \vec{\sigma} \cdot \hat{x}$, exactly as we want.

It is remarkable that even in the presence of gauge field, there is the operator

$$\Gamma_0(\vec{A}^L) = \frac{1}{2}[\Gamma(\vec{A}^L), \Gamma'(\vec{A}^L)]_+ \qquad (8.48)$$

which is in the centre of $\mathcal{A}$. It assumes the role of $\vec{J}^2$ in the presence of $\vec{A}^L$. In the continuum limit, it has the following meaning. With $D(A^L)$ denoting the Dirac operator for gauge field A^L, ($D(0)$ being the D of (8.30)), $sign\,(D(A^L))$ and $\Gamma_2(A^L)$ generate a Clifford algebra in that limit and the Hilbert space splits into a direct sum of subspaces, each carrying its IRR. $\Gamma_0(A^L)$ is a label for these subspaces.

8.4   The Basic Instanton Coupling

The instanton sectors on S^2 correspond to $U(1)$ bundles thereon. The connection on these bundles is not unique. Those with maximum symmetry have a particular simplicity and are therefore important for analysis.

In a similar way, on S_F^2, there are projective modules which in the algebraic approach substitute for sections of bundles [34] [78] [81](see chapter 5 and 6). There are particular connections on these modules with maximum symmetry and simplicity. In this section we build the Ginsparg-Wilson system for such connections. The Dirac operator then is also simple. It has zero modes which are responsible for the axial anomaly. Their presence will also be shown by simple reasoning.

To build the projective module for Chern number $2T$, $T > 0$, we follow chapters 6 and 7 and introduce $\mathbb{C}^{2T+1}$ carrying the angular momentum T representation of $SU(2)$. Let T_α, $\alpha = 1, 2, 3$ be the angular momentum operators in this representation with standard commutation relations. Let $Mat(2L + 1)^{2T+1} \equiv Mat(2L + 1) \otimes \mathbb{C}^{2T+1}$. We let $P^{(L+T)}$ be the projector coupling left angular momentum operators $\vec{L}^L$ with $\vec{T}$ to produce maximum angular momentum $L + T$. Then the projective module $P^{(L+T)} Mat(2L + 1)^{2T+1}$ is the fuzzy analogue of sections of $U(1)$ bundles on S^2 with Chern number $2T > 0$ [81]. If instead we couple $\vec{L}^L$ and $\vec{T}$ to produce the least angular momentum $(L - T)$ using the projector $P^{(L-T)}$, $P^{(L-T)} Mat(2L + 1)^{2T+1}$ corresponds to Chern number $-2T$. (We assume that $L \geq T$).

We go about as follows to set up the Ginsparg-Wilson system. For Γ, we now choose

$$\Gamma^\pm = \frac{\vec{\sigma} \cdot (\vec{L}^L + \vec{T}) + 1/2}{L \pm T + 1/2}. \tag{8.49}$$

The domain of $\Gamma^\pm$ is $P^{(L\pm T)} Mat(2L + 1)^{2T+1} \otimes \mathbb{C}^2$ with σ acting on $\mathbb{C}^2$. On this module, $(\vec{L}^L + \vec{T})^2 = (L \pm T)(L \pm T + 1)$ and $(\Gamma^\pm)^2 = \mathbb{1}$.

As for Γ', we choose it to be the same as in eq.(8.29).

$\Gamma^\pm$ and Γ' generate the new Ginsparg-Wilson system. The operators Γ_λ are defined as before as also the new Dirac operator $D^{(L\pm T)} = \frac{2}{a}\Gamma_1$. With $T > 0$ it is convenient to choose

$$a = \frac{1}{\sqrt{(L + \frac{1}{2})(L \pm T + \frac{1}{2})}} \tag{8.50}$$

for the two cases.

8.4.1   *Mixing of Spin and Isospin*

The total angular momentum $\vec{J}$ which commutes with $P^{(L\pm T)}$ and hence acts on $P^{(L\pm T)}Mat(2L+1)\otimes\mathbb{C}^2$ is not $\vec{L}^L-\vec{L}^R+\vec{\sigma}/2$, but $\vec{L}^L+\vec{T}-\vec{L}^R+\vec{\sigma}/2$. The addition of $\vec{T}$ here is the algebraic analogue of the 'mixing of spin and isospin' [70] as remarked in chapter 7. Such a term is essential in $\vec{J}$ since $\vec{L}^L - \vec{L}^R + \vec{\sigma}/2$, not commuting with $P^{(L\pm T)}$, would not preserve the modules.

8.4.2   *The Spectrum of the Dirac operator*

The spectrum of Γ_1 and $D^{(L\pm T)}$ can be derived simply by angular momentum addition, confirming the results of section 2. On the $P^{(L\pm T)}Mat(2L+1)^{2T+1}$ modules, $(\vec{L}^L + \vec{T})^2$ has the fixed values $(L\pm T)(L\pm T + 1)$, and

$$(\Gamma_1)^2 = \frac{1}{(2(L\pm T) + 1)(2L+1)}((\vec{L}^L + \vec{T} - \vec{L}^R + \frac{1}{2}\vec{\sigma})^2 + \frac{1}{4} - T^2), \quad (8.51)$$

$$\Gamma^{\pm} = \frac{(\vec{L}^L + \vec{T} + \frac{1}{2}\vec{\sigma})^2 - (L\pm T)(L\pm T + 1) - \frac{1}{4}}{(L\pm T) + \frac{1}{2}}, \quad (8.52)$$

$$\Gamma' = \frac{(-\vec{L}^R + \frac{1}{2}\vec{\sigma})^2 - L(L+1) - \frac{1}{4}}{L + \frac{1}{2}}. \quad (8.53)$$

Comparing (8.51) with (8.10) we see that the 'total angular momentum' $(\vec{J})^2 = (\vec{L}^L + \vec{T} - \vec{L}^R + \frac{1}{2}\vec{\sigma})^2$ is linearly related to $\Gamma_0 = \frac{1}{2}[\Gamma^{\pm}, \Gamma']_+$. The eigenvalues $(\gamma_1)^2$ of $(\Gamma_1)^2$ are determined by those of $(\vec{J})^2$, call them $j(j+1)$.

For $j = j_{max} = L \pm T + L + \frac{1}{2}$ we have $(\Gamma_1)^2 = 1$, so this is V_{+1}, and the degeneracy is $2j_{max} + 1 = 2(2L\pm T + 1)$. The maximum value of j can be achieved only if

$$(\vec{L}^L+\vec{T}+\frac{1}{2}\vec{\sigma})^2 = (L\pm T+\frac{1}{2})(L\pm T+\frac{3}{2}), \quad (-\vec{L}^R + \frac{1}{2}\vec{\sigma})^2 = (L+\frac{1}{2})(L+\frac{3}{2}). \quad (8.54)$$

Replacing these values in (8.52,8.53) we see that on V_{+1} we have $\gamma_1 = 1$, and $\Gamma_2 = 0$.

The case $T = 0$ has been treated before [7][81][99].So we here assume that $T > 0$. In that case, for either module $j_{min} = T - \frac{1}{2}$, which gives an eigenvalue $(\gamma_1)^2 = 0$ with degeneracy $2T$; we are in V_{-1}, the space of the zero modes. To realize this minimum value of j we must have

$$(\vec{L}^L + \vec{T} + \frac{1}{2}\vec{\sigma})^2 = (L \pm T \mp \frac{1}{2})(L \pm T \mp \frac{1}{2} + 1), \quad (8.55)$$

$$(-\vec{L}^R + \frac{1}{2}\vec{\sigma})^2 = (L \pm \frac{1}{2})(L \pm \frac{1}{2} + 1). \quad (8.56)$$

Replacing these values in (8.52, 8.53) we find that on the corresponding eigenstates $\Gamma_2 = \mp 1$: they are all either chiral left or chiral right. These are the results needed by continuum index theory and axial anomaly.

For $j_{min} < j < j_{max}$, that is on V, we have $0 < (\gamma_1)^2 < 1$, and by (8.12), $\Gamma_2 \neq 0$. Since $[\Gamma_1, \Gamma_2]_+ = 0$, to each state ψ such that $\Gamma_1 \psi = \gamma_1 \psi$ corresponds a state $\psi' = \Gamma_2 \psi$ such that $\Gamma_1 \psi' = -\gamma_1 \psi'$.

For any value of j we can write $j = n + T - \frac{1}{2}$ with $n = 0, 1, ..., 2L + 1$ when the projector is $P^{(L+T)}$, and $n = 0, 1, ..., 2(L - T) + 1$ when the projector is $P^{(L-T)}$, while correspondingly,

$$(\gamma_1)^2 = \frac{n(n + 2T)}{(2(L \pm T) + 1)(2L + 1)} . \tag{8.57}$$

With the choice (8.50) for a this gives for the squared Dirac operator the eigenvalues $\rho^2 = n(n + 2T)$. This spectrum agrees *exactly* with what one finds in the continuum [100], except at the top value of n. Such a result is true also for $T = 0$ [99][81]. For the top value of n, $\Gamma_2 = 0$, and we get only the eigenvalue $\gamma_1 = 1$, whereas in the continuum, $\Gamma_2 \neq 0$ and both eigenvalues $\gamma_1 = \pm 1$ occur. This result [99][81], valid also for $T = 0$, has been known for a long time.

Finally, we can check that summing the degeneracies of the eigenvalues we have found, we get exactly the dimension of the corresponding module. In fact:

$$2T + 2 \sum_{n=1}^{2L} \left(2(n + T - \frac{1}{2}) + 1 \right) + 2(2L + T + 1)$$
$$= 2(2L + 1)(2(L + T) + 1) ,$$
$$2T + 2 \sum_{n=1}^{2(L-T)} \left(2(n + T - \frac{1}{2}) + 1 \right) + 2(2L - T + 1)$$
$$= 2(2L + 1)(2(L - T) + 1) . \tag{8.58}$$

We show below that the axial anomaly on S_F^2 is stable against perturbations compatible with the chiral properties of the Dirac operator, and is hence a 'topological' invariant.

8.5 Gauging the Dirac Operator in Instanton Sectors

The operator $\vec{\mathcal{L}} + \vec{T}$ commutes with $P^{(L \pm T)}$ and hence preserves the projective modules. It is important to preserve this feature on gauging as well.

So the gauge field $\vec{A}^L$ is taken to be a function of $\vec{L}^L + \vec{T}$ (which remains bounded as $L \to \infty$). For $L \to \infty$, it becomes a function of x. The limiting transversality of $\vec{T} + \vec{A}^L$ can be guaranteed by imposing the condition

$$(\vec{L}^L + \vec{T} + \vec{A}^L)^2 = (\vec{L}^L + \vec{T})^2 = (L \pm T)(L \pm T + 1) , \qquad (8.59)$$

which generalizes (8.42).

We can now construct the Ginsparg-Wilson system using

$$\Gamma(A^L) = \frac{\sigma \cdot (\vec{L}^L + \vec{T} + \vec{A}^L) + 1/2}{|\sigma \cdot (\vec{L}^L + \vec{T} + \vec{A}^L) + 1/2|} \qquad (8.60)$$

and the Γ' of (8.29), $\Gamma(0)$ being Γ of (8.49). $\sigma \cdot (\vec{L}^L + \vec{T}) + 1/2$ has no zero modes, and therefore (8.60) is well-defined for generic $\vec{A}^L$. We can now use section 2 to construct the Dirac theory.

We have a continuous number of Ginsparg-Wilson algebras labeled by $\vec{A}^L$. For each, (8.21) holds:

$$\mathrm{Tr}\,\Gamma_2(A^L) = n(A^L) . \qquad (8.61)$$

Here as $n(A^L) \in \mathbb{Z}$, it is in fact a constant by continuity. The index of the Dirac operator and the global $U(1)_A$ axial anomaly implied by (8.61) are thus independent of $\vec{A}^L$ as previously indicated. [See Fujikawa [96] and [97] for the connection of (8.61) to the global axial anomaly.]

The expansions (8.45-8.47) are easily extended to the instanton sectors, and imply the desired continuum limit of $D^{(L\pm T)}(\vec{A}^L)$ and chirality $\Gamma_2(\vec{A}^L)$:

$$D^{(L\pm T)}(\vec{A}^L) \ \to \ \vec{\sigma} \cdot (\vec{\mathcal{L}} + \vec{T} + \vec{A}) + 1 ,$$
$$\Gamma_2(A^L) \ \to \ \vec{\sigma} \cdot \hat{x} . \qquad (8.62)$$

Chirality is thus independent of the gauge field in the limiting case, but not otherwise.

8.6   Further Remarks on the Axial Anomaly

The local form of $U(1)_A$-anomaly has not been treated in the present approach. (See however [63][100][101].) As for gauge anomalies, the central and familiar problem is that noncommutative algebras allow gauging only by the particular groups $U(N)$, and that too by their particular representations (see chapter 7). This is so in a naive approach. There are clever methods to overcome this problem on the Moyal planes [102] using the Seiberg-Witten map [103], but they fail for the fuzzy spaces. Thus gauge anomalies can be studied for fuzzy spaces only in a very limited manner, but even this is yet to be done. More elaborate issues like anomaly cancellation in a fuzzy version of the standard model have to wait till the above mentioned problems are solved.

Chapter 9

Fuzzy Supersymmetry

Another important feature we encounter in studying fuzzy discretizations is their ability to preserve supersymmetry (SUSY) exactly: They allow the formulation of regularized and exactly supersymmetric field theories. It is very difficult to formulate models with exact SUSY in conventional lattice discretizations. At least for this reason, fuzzy supersymmetric spaces merit careful study.

The original idea of a fuzzy supersphere is due to Grosse et al.[7, 8]. A slightly different approach for its construction, which is closer to ours, is given in [122].

We start this chapter describing the supersphere $S^{(2,2)}$ and its fuzzy version $S_F^{(2,2)}$. Although the mathematical structure underlying the formulation of the supersphere is a generalization of that of the 2-sphere, it is not widely known. Therefore, we here collect the necessary information on representation theory and basic properties of Lie superalgebras $osp(2,1)$ and $osp(2,2)$ and their corresponding supergroups $OSp(2,1)$ and $OSp(2,2)$ from sections 9.1 to 9.3: they underlie the construction of $S^{(2,2)}$ and consequently that of $S_F^{(2,2)}$ as discussed in these sections.

In section 9.4, the construction of generalized coherent states is extended to the supergroup $OSp(2,1)$.

In section 9.5, we outline the SUSY action of Grosse et al. [7] on $S^{(2,2)}$. It is a quadratic action in scalar and spinor fields. It is the simplest SUSY action one can formulate and is closest to the quadratic scalar field action on S^2. We then discuss its fuzzy version. The latter has exact SUSY.

Following three sections discuss the construction and differential geometric properties of an associative $*$-product of functions on $S_F^{(2,2)}$ and on "sections of bundles" on $S_F^{(2,2)}$.

We conclude the chapter by a brief discussion on construction of nonlinear sigma models on $S_F^{(2,2)}$.

Our discussion in this chapter follows and expands upon [9].

9.1 $osp(2,1)$ and $osp(2,2)$ Superalgebras and their Representations

Here we review some of the basic features regarding the Lie superalgebras $osp(2,1)$ and $osp(2,2)$. For detailed discussions, the reader is refered to the references [108–112].

The Lie superalgebras $osp(2,1)$ and $osp(2,2)$ can be defined in terms of 3×3 matrices acting on $\mathbb{C}^3$. The vector space $\mathbb{C}^3$ is graded: it is to be regarded as $\mathbb{C}^2 \oplus \mathbb{C}^1$ where $\mathbb{C}^2$ is the even- and $\mathbb{C}^1$ is the odd-subspace. As $\mathbb{C}^3$ is so graded, it is denoted by $\mathbb{C}(2,1)$ while linear operators on $\mathbb{C}(2,1)$ are denoted by $Mat(2,1)$. ($\mathbb{C}(2,1)$ is to be distinguished from the superspace $\mathcal{C}^{(2,1)}$ which will appear in section (9.3). By convention, the above $\mathbb{C}^2$ and $\mathbb{C}^1$ are embedded in $\mathbb{C}^3$ as follows:

$$\mathbb{C}^2 = \{(\xi_1, \xi_2, 0) : \xi_i \in \mathbb{C}\} \subset \mathbb{C}^{(2,1)},$$
$$\mathbb{C}^1 = \{(0, 0, \eta) : \eta \in \mathbb{C}\} \subset \mathbb{C}^{(2,1)}. \tag{9.1}$$

The grade of $\mathbb{C}^2$ is 0 (mod 2) and that of $\mathbb{C}^1$ is 1 (mod 2). The grading of $\mathbb{C}(2,1)$ induces a grading of $Mat(2,1)$. A linear operator $L \in Mat(2,1)$ has grade $|L| = 0$ (mod 2) or is "even" if it does not change the grade of underlying vectors of definite grade. Such an L is block-diagonal:

$$L = \begin{pmatrix} \ell_1 & \ell_2 & 0 \\ \ell_3 & \ell_4 & 0 \\ 0 & 0 & \ell \end{pmatrix} \quad \ell_i, \ell \in \mathbb{C} \; if \; |L| = 0 \pmod 2. \tag{9.2}$$

If L instead changes the grade of an underlying vector of definite grade by 1 (mod 2) unit, its grade is $|L| = 1$ (mod 2) or it is "odd". Such an L is off-diagonal:

$$L = \begin{pmatrix} 0 & 0 & s_1 \\ 0 & 0 & s_2 \\ t_1 & t_2 & 0 \end{pmatrix} \quad s_i, t_i \in \mathbb{C} \quad if \quad |L| = 1 \pmod 2. \tag{9.3}$$

A generic element of $\mathbb{C}(2,1)$ and $Mat(2,1)$ will be a sum of elements of both grades and will have no definite grade.

If $M, N \in Mat(2,1)$ have definite grades $|M|$, $|N|$ their graded Lie bracket $[M, N\}$ is defined by

$$[M, N\} = MN - (-1)^{|M||N|} NM. \tag{9.4}$$

The even part of $osp(2,1)$ is the Lie algebra $su(2)$ for which $\mathbb{C}^2$ has spin $\frac{1}{2}$ and $\mathbb{C}^1$ has spin 0. $su(2)$ has the usual basis

$$\Lambda_i^{(\frac{1}{2})} = \frac{1}{2} \begin{pmatrix} \sigma_i & 0 \\ 0 & 0 \end{pmatrix}, \sigma_i = \text{Pauli matrices}. \tag{9.5}$$

The superscript $\frac{1}{2}$ here denotes this representation: irreducible representations of $osp(2,1)$ are labelled by the highest angular momentum.

$osp(2,1)$ has two more generators $\Lambda_\alpha^{(\frac{1}{2})}$ ($\alpha = 4,5$) in its basis:

$$\Lambda_4^{(\frac{1}{2})} = \frac{1}{2} \begin{pmatrix} 0 & 0 & -1 \\ 0 & 0 & 0 \\ 0 & -1 & 0 \end{pmatrix}, \quad \Lambda_5^{(\frac{1}{2})} = \frac{1}{2} \begin{pmatrix} 0 & 0 & 0 \\ 0 & 0 & -1 \\ 1 & 0 & 0 \end{pmatrix}. \tag{9.6}$$

The full $osp(2,1)$ superalgebra is defined by the graded commutators

$$[\Lambda_i^{(\frac{1}{2})}, \Lambda_j^{(\frac{1}{2})}] = i\epsilon_{ijk}\Lambda_k^{(\frac{1}{2})}, [\Lambda_i^{(\frac{1}{2})}, \Lambda_\alpha^{(\frac{1}{2})}] = \frac{1}{2}(\sigma_i)_{\beta\alpha}\Lambda_\beta^{(\frac{1}{2})},$$

$$\{\Lambda_\alpha^{(\frac{1}{2})}, \Lambda_\beta^{(\frac{1}{2})}\} = \frac{1}{2}(C\sigma_i)_{\alpha\beta}\Lambda_i^{(\frac{1}{2})}, \tag{9.7}$$

where $C_{\alpha\beta} = -C_{\beta\alpha}$ is the Levi-Civita symbol with $C_{45} = 1$. (Here the rows and columns of σ_i and C are being labeled by $4,5$).

The abstract $osp(2,1)$ Lie superalgebra has basis $\Lambda_i, \Lambda_\alpha$ ($i = 1,2,3, \alpha = 4,5$) with graded commutators obtained from (9.7) by dropping the superscript $\frac{1}{2}$:

$$[\Lambda_i, \Lambda_j] = i\epsilon_{ijk}\Lambda_k, \quad [\Lambda_i, \Lambda_\alpha] = \frac{1}{2}(\sigma_i)_{\beta\alpha}\Lambda_\beta, \quad \{\Lambda_\alpha, \Lambda_\beta\} = \frac{1}{2}(C\sigma_i)_{\alpha\beta}\Lambda_i. \tag{9.8}$$

Thus Λ_α transforms like an $su(2)$ spinor.

The Lie algebra $su(2)$ is isomorphic to the Lie algebra $osp(2)$ of the ortho-symplectic group $OSp(2)$. The above graded Lie algebra has in addition one spinor in its basis. For this reason, it is denoted by $osp(2,1)$.

In customary Lie algebra theory, compactness of the underlying group is reflected in the adjointness properties of its Lie algebra elements. Thus these Lie algebras allow a star $*$ or adjoint operation $\dagger$ and their elements are invariant under $\dagger$ (in the convention of physicists) if the underlying group is compact. As $\dagger$ complex conjugates complex numbers, the Lie algebras of compact Lie groups are real as vector spaces: they are real Lie algebras.

In graded Lie algebras, the operation $\dagger$ is replaced by the grade adjoint (or grade star) operation $\ddagger$. Its relation to the properties of the underlying supergroup will be indicated later. The properties and definition of $\ddagger$ are as follows.

First, we note that the grade adjoint of an even (odd) element is even (odd). Next, one has $(A^{\ddagger})^{\ddagger} = (-1)^{|A|}A$ for an even or odd (that is homogeneous) element A of degree $|A|$ (mod 2), or equally well, integer (mod 2). (So, depending on $|A|$, $|A|$ itself can be taken to be 0 or 1.) Thus, it is the usual † on the even part, while on an odd element A, it squares to -1. Further $(AB)^{\ddagger} = (-1)^{|A||B|}B^{\ddagger}A^{\ddagger}$ so that, $[A, B]^{\ddagger} = (-1)^{|A||B|}[B^{\ddagger}, A^{\ddagger}]$ for homogeneous elements A, B.

Henceforth, we will denote the degree of a (which may be a Lie superalgebra element, a linear operator or an index) by $|a|$(mod 2), $|a|$ denoting any integer in its equivalence class $\langle |a| + 2n : n \in \mathbb{Z} \rangle$.

The basis elements of the $osp(2, 1)$ (and $osp(2, 2)$, see later) graded Lie algebras are taken to fulfill certain "reality" properties implemented by $\ddagger$. For the generators of $osp(2, 1)$, these are given by

$$\Lambda_i^{\ddagger} = \Lambda_i^{\dagger} = \Lambda_i, \quad \Lambda_{\alpha}^{\ddagger} = -\sum_{\beta=4,5} C_{\alpha\beta}\Lambda_{\beta} \quad \alpha = 4, 5. \tag{9.9}$$

Let V be a graded vector space V so that $V = V_0 \oplus V_1$ where V_0 and V_1 are even and odd subspaces [112]. In a (grade star) representation of a graded Lie algebra on V, V_0 and V_1 are invariant under the even elements of the graded Lie algebra while its odd elements map one to the other.

This representation becomes a grade-$*$ representation if the following is also true. Let us assume that V is endowed with the inner product $\langle u|v \rangle$ for all $u, v \in V$. Now if L is a linear operator acting on V, then the grade adjoint of L is defined by

$$\langle L^{\ddagger}u|v \rangle = (-1)^{|u||L|}\langle u|Lv \rangle \tag{9.10}$$

for homogeneous elements u, L. In a basis adapted to the above decomposition of V, a generic L has the matrix representation

$$M_L = \begin{pmatrix} \alpha_1 & \alpha_2 \\ \alpha_3 & \alpha_4 \end{pmatrix} = M_0 + M_1, \quad M_0 = \begin{pmatrix} \alpha_1 & 0 \\ 0 & \alpha_4 \end{pmatrix}, \quad M_1 = \begin{pmatrix} 0 & \alpha_2 \\ \alpha_3 & 0 \end{pmatrix} \tag{9.11}$$

where M_0 and M_1 are the even and odd parts of M_L. The formula for $\ddagger$ is then

$$M_L^{\ddagger} = \begin{pmatrix} \alpha_1^{\dagger} & -\alpha_3^{\dagger} \\ \alpha_2^{\dagger} & \alpha_4^{\dagger} \end{pmatrix}, \tag{9.12}$$

$\alpha_i^{\dagger}$ being matrix adjoint of α_i.

Then in a grade-$*$ representation, the image of $L^{\ddagger}$ is $M_L^{\ddagger}$.

We note that the supertrace str of M_L is by definition

$$strM_L = Tr\alpha_1 - Tr\alpha_4. \tag{9.13}$$

The irreducible representations of $osp(2,1)$ are characterized by an integer or half-integer non-negative quantum number $J_{osp(2,1)}$ called superspin. From the point of view of the irreducible representations of $su(2)$, the superspin $J_{osp(2,1)}$ representation has the decomposition

$$J_{osp(2,1)} = J_{su(2)} \oplus \left(J - \frac{1}{2} \right)_{su(2)}, \qquad (9.14)$$

where $J_{su(2)}$ is the $su(2)$ representation for angular momentum J. All these are grade-$*$ representations : the relations (9.9) are preserved in the representation.

The fundamental and adjoint representations of $osp(2,1)$ correspond to $J_{osp(2,1)} = \frac{1}{2}$ and $J_{osp(2,1)} = 1$ respectively, being 3 and 5 dimensional. The quadratic Casimir operator is

$$K_2^{osp(2,1)} = \Lambda_i \Lambda_i + C_{\alpha\beta} \Lambda_\alpha \Lambda_\beta. \qquad (9.15)$$

It has eigenvalues $J(J + \frac{1}{2})$.

It is also worthwhile to make the following technical remark. The superspin multiplets in $J_{osp(2,1)}$ representation may be denoted by $|J_{osp(2,1)}$, $J_{su(2)}, J_3\rangle$, and $|J_{osp(2,1)}, \left(J - \frac{1}{2} \right)_{su(2)}, J_3\rangle$. One of the multiplets generates the even and the other generates the odd subspace of the representation space. Although, this grade can be arbitrarily assigned, the choice consistent with the reality conditions we have chosen in (9.9) and the definition of grade adjoint operation in (9.10) fixes the multiplet $|J_{osp(2,1)}, J_{su(2)}, J_3\rangle$ to be of even degree while $|J_{osp(2,1)}, \left(J - \frac{1}{2} \right)_{su(2)}, J_3\rangle$ is odd.

The $osp(2,2)$ superalgebra can be defined by introducing an even generator Λ_8 commuting with the Λ_i and odd generators Λ_α with $\alpha = 6,7$ in addition to the already existing ones for $osp(2,1)$. The graded commutation relations for $osp(2,2)$ are then

$$[\Lambda_i, \Lambda_j] = i\epsilon_{ijk}\Lambda_k, \quad [\Lambda_i, \Lambda_\alpha] = \frac{1}{2}(\tilde{\sigma}_i)_{\beta\alpha}\Lambda_\beta, \quad [\Lambda_i, \Lambda_8] = 0,$$

$$[\Lambda_8, \Lambda_\alpha] = \tilde{\varepsilon}_{\alpha\beta}\Lambda_\beta, \quad \{\Lambda_\alpha, \Lambda_\beta\} = \frac{1}{2}(\tilde{C}\tilde{\sigma}_i)_{\alpha\beta}\Lambda_i + \frac{1}{4}(\tilde{\varepsilon}\tilde{C})_{\alpha\beta}\Lambda_8, \qquad (9.16)$$

where $i, j = 1, 2, 3$ and $\alpha, \beta = 4, 5, 6, 7$. In above we have used the matrices

$$\tilde{\sigma}_i = \begin{pmatrix} \sigma_i & 0 \\ 0 & \sigma_i \end{pmatrix}, \quad \tilde{C} = \begin{pmatrix} C & 0 \\ 0 & -C \end{pmatrix}, \quad \tilde{\varepsilon} = \begin{pmatrix} 0 & I_{2\times2} \\ I_{2\times2} & 0 \end{pmatrix}. \qquad (9.17)$$

Their matrix elements are indexed by $4, \ldots, 7$.

In addition to (9.9), the new generators satisfy the "reality" conditions

$$\Lambda_\alpha^\dagger = -\sum_{\beta=6,7} \tilde{C}_{\alpha\beta}\Lambda_\beta, \quad \alpha = 6, 7, \qquad \Lambda_8^\ddagger = \Lambda_8^\dagger = \Lambda_8. \qquad (9.18)$$

So we can write the $osp(2,2)$ reality conditions for all α as $\Lambda_\alpha^\ddagger = -\tilde{C}_{\alpha\beta}\Lambda_\beta$.

Irreducible representations of $osp(2,2)$ fall into two categories, namely the typical and non-typical ones. Both are grade $*$-representations which preserve the reality conditions (9.9) and (9.18). Typical ones are reducible with respect to the $osp(2,1)$ superalgebra (except for the trivial representation) whereas non-typical ones are irreducible. Typical representations are labeled by an integer or half integer non-negative number $J_{osp(2,2)}$, called $osp(2,2)$ superspin and the maximum eigenvalue k of Λ_8 in that IRR. They can be denoted by $(J_{osp(2,2)}, k)$. Independently of k, these have the $osp(2,1)$ content $J_{osp(2,1)} \oplus (J - \frac{1}{2})_{osp(2,1)}$ for $J_{osp(2,2)} \geq \frac{1}{2}$ while $(0)_{osp(2,2)}$ contains just $(0)_{osp(2,1)}$. Hence on restriction to $su(2)$,

$$\left(J_{osp(2,2)}, k\right) \rightarrow$$
$$\begin{cases} J_{su(2)} \oplus \left(J - \frac{1}{2}\right)_{su(2)} \oplus \left(J - \frac{1}{2}\right)_{su(2)} \oplus (J - 1)_{su(2)}\,, & J_{osp(2,2)} \geq 1\,; \\ (\frac{1}{2})_{su(2)} + (0)_{su(2)} + (0)_{su(2)}\,, & J_{osp(2,2)} = \frac{1}{2}\,. \end{cases}$$
$$(9.19)$$

$osp(2,2)$ has the quadratic Casimir operator

$$K_2^{osp(2,2)} = \Lambda_i\Lambda_i + \tilde{C}_{\alpha\beta}\Lambda_\alpha\Lambda_\beta - \frac{1}{4}\Lambda_8^2$$

$$= K_2^{osp(2,1)} - \left(\sum_{\alpha,\beta=6,7} -\tilde{C}_{\alpha\beta}\Lambda_\alpha\Lambda_\beta + \frac{1}{4}\Lambda_8^2\right)\,. \qquad (9.20)$$

It has also a cubic Casimir operator [108, 113]. We do not show it here , as we will not use it.

Note that since all the generators of $osp(2,1)$ commute with $K_2^{osp(2,2)}$ and $K_2^{osp(2,1)}$, they also commute with

$$K_2^{osp(2,1)} - K_2^{osp(2,2)} = -\sum_{\alpha,\beta=6,7} \tilde{C}_{\alpha\beta}\Lambda_\alpha\Lambda_\beta + \frac{1}{4}\Lambda_8^2\,. \qquad (9.21)$$

The $osp(2,2)$ Casimir $K_2^{osp(2,2)}$ vanishes on non-typical representations:

$$K_2^{osp(2,2)}\Big|_{nontypical} = 0\,. \qquad (9.22)$$

The substitutions

$$\Lambda_i \rightarrow \Lambda_i\,, \quad \Lambda_\alpha \rightarrow \Lambda_\alpha\,, \quad \alpha = 4,5; \quad \Lambda_\alpha \rightarrow -\Lambda_\alpha\,, \quad \alpha = 6,7; \quad \Lambda_8 \rightarrow -\Lambda_8$$
$$(9.23)$$

define an automorphism of $osp(2,2)$. This automorphism changes the irreducible representation $(J_{osp(2,2)}, k)$ into an inequivalent one $(J_{osp(2,2)}, -k)$

(except for the trivial representation with $J = 0$), while preserving the reality conditions given in (9.9) and (9.18) [109]. In the nontypical case, we discriminate between these two representations associated with $J_{osp(2,1)}$ as follows: For $J > 0$, $J_{osp(2,2)+}$ will denote the representation in which the eigenvalue of the representative of Λ_8 on vectors with angular momentum J is positive and $J_{osp(2,2)-}$ will denote its partner where this eigenvalue is negative. (This eigenvalue is zero only in the trivial representation with $J = 0$.) Here while considering nontypical IRR's we concentrate on $J_{osp(2,2)+}$. The results for $J_{osp(2,2)-}$ are similar and will be occasionally indicated.

Another important result in this regard is that every non-typical representation $J_{osp(2,2)\pm}$ of $osp(2,2)$, is at the same time an irreducible representation of $osp(2,1)$ with superspin $J_{osp(2,1)}$. For this reason the $osp(2,2)$ generators $\Lambda_{6,7,8}$ can be nonlinearly realized in terms of the $osp(2,1)$ generators. Repercussions of this result will be seen later on.

Below we list some of the well-known results and standard notations that are used throughout the text. The fundamental representation of $osp(2,2)$ is non-typical and we concentrate on the one given by $J_{osp(2,2)+} = (\frac{1}{2})_{osp(2,2)+}$. It is generated by the (3×3) supertraceless matrices $\Lambda_a^{(\frac{1}{2})}$ satisfying the "reality" conditions of (9.9) and (9.18):

$$\Lambda_i^{(\frac{1}{2})} = \frac{1}{2}\begin{pmatrix} \sigma_i & 0 \\ 0 & 0 \end{pmatrix}, \quad \Lambda_4^{(\frac{1}{2})} = \frac{1}{2}\begin{pmatrix} 0 & \xi \\ \eta^T & 0 \end{pmatrix}, \quad \Lambda_5^{(\frac{1}{2})} = \frac{1}{2}\begin{pmatrix} 0 & \eta \\ -\xi^T & 0 \end{pmatrix},$$

$$\Lambda_6^{(\frac{1}{2})} = \frac{1}{2}\begin{pmatrix} 0 & -\xi \\ \eta^T & 0 \end{pmatrix}, \quad \Lambda_7^{(\frac{1}{2})} = \frac{1}{2}\begin{pmatrix} 0 & -\eta \\ -\xi^T & 0 \end{pmatrix}, \quad \Lambda_8^{(\frac{1}{2})} = \begin{pmatrix} I_{2\times2} & 0 \\ 0 & 2 \end{pmatrix},$$

$$(9.24)$$

where

$$\xi = \begin{pmatrix} -1 \\ 0 \end{pmatrix} \quad \text{and} \quad \eta = \begin{pmatrix} 0 \\ -1 \end{pmatrix}. \tag{9.25}$$

These generators satisfy

$$\Lambda_a^{(\frac{1}{2})}\Lambda_b^{(\frac{1}{2})} = S_{ab}\mathbf{1} + \frac{1}{2}\left(d_{abc} + if_{abc}\right)\Lambda_c^{(\frac{1}{2})} \quad (a,b,c = 1,2,\ldots 8). \tag{9.26}$$

It is possible to write

$$S_{ab} = str\left(\Lambda_a^{(\frac{1}{2})}\Lambda_b^{(\frac{1}{2})}\right), f_{abc} = str\left(-i[\Lambda_a^{(\frac{1}{2})}, \Lambda_b^{(\frac{1}{2})}\}\Lambda_c^{(\frac{1}{2})}\right),$$

$$d_{abc} = str\left(\{\Lambda_a^{(\frac{1}{2})}, \Lambda_b^{(\frac{1}{2})}]\Lambda_c^{(\frac{1}{2})}\right). \tag{9.27}$$

Here $a = i = 1, 2, 3$, and $a = 8$ label the even generators whereas $a = \alpha = 4, 5, 6, 7$ label the odd generators.

In above $[A, B], \{A, B\}$ denote the graded commutator and the graded anticommutator respectively. The former is already defined, while the latter is given by $\{A, B\} = AB + (-1)^{|A||B|}BA$ for homogeneous elements A and B.

S_{ab} defines the invariant metric of the Lie superalgebra $osp(2, 2)$. In their block diagonal form, S and its inverse read

$$S = \begin{pmatrix} \frac{1}{2}I & & \\ & -\frac{1}{2}\tilde{C} & \\ & & -2 \end{pmatrix}_{8\times 8} \quad , \quad S^{-1} = \begin{pmatrix} 2I & & \\ & 2\tilde{C} & \\ & & -\frac{1}{2} \end{pmatrix}_{8\times 8} . \tag{9.28}$$

The explicit values of the structure constants f_{abc} can be read from (9.17), since $[\Lambda_a, \Lambda_b\} = if_{abc}\Lambda_c$. Those of d_{abc} are as follows[*]:

$$d_{ij8} = -\frac{1}{2}\delta_{ij}, \quad d_{\alpha\beta 8} = \frac{3}{4}\tilde{C}_{\alpha\beta}, \quad d_{\alpha 8\beta} = 3\delta_{\alpha\beta}, \quad d_{i8j} = 2\delta_{ij},$$

$$d_{\alpha\beta i} = -\frac{1}{2}(\tilde{\varepsilon}\tilde{C}\tilde{\sigma}_i)_{\alpha\beta}, \quad d_{i\alpha\beta} = -\frac{1}{2}(\tilde{\varepsilon}\tilde{\sigma}_i)_{\beta\alpha}, \quad d_{888} = 6. \tag{9.29}$$

We close this subsection with a final remark. Discussion in the subsequent sections will involve the use of linear operators acting on the adjoint representation of $osp(2, 2)$. These are linear operators $\widehat{\mathcal{Q}}$ acting on Λ_a according to $\widehat{\mathcal{Q}}\Lambda_a = \Lambda_b \mathcal{Q}_{ba}$, $\mathcal{Q}$ being the matrix representation of $\widehat{\mathcal{Q}}$. They are graded because Λ_a's are, and hence the linear operators on the adjoint representation are graded. The degree (or grade) of a matrix $\mathcal{Q}$ with only the nonzero entry $\mathcal{Q}_{ab}$ is $(|\Lambda_a| + |\Lambda_b|)\,(mod\,2) \equiv (|a| + |b|)\,(mod\,2)$. The grade star operation on $\widehat{\mathcal{Q}}$ now follows from the sesquilinear form

$$\left(\alpha = \alpha_a\Lambda_A, \, \beta = \beta_b\Lambda_b\right) = \bar{\alpha}_a S_{ab}^{-1}\beta_b, \qquad \alpha_a, \beta_b \in \mathbb{C} \tag{9.30}$$

and is given by

$$(\widehat{\mathcal{Q}}^{\ddagger}\alpha, \beta) = (-1)^{|\alpha||\widehat{\mathcal{Q}}|}(\alpha, \widehat{\mathcal{Q}}\beta). \tag{9.31}$$

9.2 Passage to Supergroups

We recollect here the passage from these superalgebras to their corresponding supergroups [112, 114]. Let $\xi \equiv (\xi_1, \cdots, \xi_8)$ be the elements of the superspace $\mathbb{R}^{(4,4)}$. Here ξ_a for $a = i = 1, 2, 3$ and $a = 8$ label the even and for $a = \alpha = 4, 5, 6, 7$ label the odd elements of a real Grassmann algebra $\mathcal{G}$. ξ_a's satisfy the graded commutation relations mutually and with the algebra elements:

$$[\xi_a, \xi_b\} = 0, \qquad [\xi_a, \Lambda_b\} = 0. \tag{9.32}$$

---

[*]The tensor d_{abc} given explicitly in (9.29) for $J_{osp(2,2)+}$ becomes $-d_{abc}$ for $J_{osp(2,2)-}$.

We assume that $\xi_i^\dagger = \xi_i, \xi_8^\dagger = \xi_8$ and $\xi_\alpha^\dagger = -\tilde{C}_{\alpha\beta}\xi_\beta$. Then $\xi_a\Lambda_a$ is grade-$*$ even:

$$(\xi_a\Lambda_a)^\ddagger = \xi_a\Lambda_a. \tag{9.33}$$

An element of $OSp(2,2)$ is given by $g = e^{i\xi_a\Lambda_a}$, while for a restricted to $a \leq 5$, g gives an element of $OSp(2,1)$. (9.33) corresponds to the usual hermiticity property of Lie algebras which yields unitary representations of the group.

9.3 On the Superspaces

9.3.1 *The Superspace $\mathcal{C}^{2,1}$ and the Noncommutative $\mathcal{C}_F^{2,1}$*

$\mathcal{C}^{2,1}$ is the $(2,1)$-dimensional superspace specified by two even and one odd element of a complex Grassmann algebra $\mathcal{G}$. Let $\mathcal{G}_0$ and $\mathcal{G}_1$ denote the even and odd subspaces of $\mathcal{G}$. We write

$$\mathcal{C}^{2,1} \equiv \{\psi \equiv (z_1, z_2, \theta)\}, \tag{9.34}$$

where $z_1, z_2 \in \mathcal{G}_0$ and $\theta \in \mathcal{G}_1$ satisfy

$$\{\theta, \bar{\theta}\} \equiv \theta\bar{\theta} + \bar{\theta}\theta = 0, \quad \theta\theta = \bar{\theta}\bar{\theta} = 0. \tag{9.35}$$

We note that under $\ddagger$ operation

$$z_i^\ddagger = z_i^\dagger = \bar{z}_i, \quad \theta^\ddagger = \bar{\theta}, \quad \bar{\theta}^\ddagger = -\theta. \tag{9.36}$$

The noncommutative $\mathcal{C}^{2,1}$, denoted by $\mathcal{C}_F^{2,1}$ hereafter, is obtained by replacing $\psi \in \mathcal{C}^{2,1}$, by $\Psi \equiv (a_1, a_2, b)$, where the operators a_i and b obey the commutation and anticommutation relations

$$[a_i\, a_j] = [a_i^\dagger\, a_j^\dagger] = 0, \quad [a_i\, a_j^\dagger] = \delta_{ij}, \quad [a_i, b] = [a_i, b^\dagger] = 0$$
$$\{b, b\} = \{b^\dagger, b^\dagger\} = 0, \quad \{b, b^\dagger\} = 1. \tag{9.37}$$

Under $\dagger$ they fulfill $a_i^\ddagger = a_i$, $(a_i^\dagger)^\ddagger = a_i$, $b^\ddagger = b^\dagger$, $(b^\dagger)^\ddagger = -b$.

Using the notation

$$(\Psi_1, \Psi_2, \Psi_0) \equiv (a_1, a_2, b), \tag{9.38}$$

the commutation relations can be more compactly expressed as

$$[\Psi_\mu, \Psi_\nu\} = [\Psi_\mu^\dagger, \Psi_\nu^\dagger\} = 0, \quad [\Psi_\mu, \Psi_\nu^\dagger\} = \delta_{\mu\nu}, \tag{9.39}$$

where $\mu = 1, 2, 0$. Ψ_μ, $\Psi_\mu^\dagger$ and the identity operator $\mathbf{1}$ span the graded Heisenberg-Weyl algebra, with $\mathbf{1}$ generating its center.

9.3.2 The Supersphere $S^{(3,2)}$ and the Noncommutative $S^{(3,2)}$

Dividing ψ by its modulus $|\psi| \equiv |z_1|^2 + |z_2|^2 + \bar{\theta}\theta$, we define $\psi' = \frac{\psi}{|\psi|} \in \mathcal{C}^{2,1} \setminus \{0\}$ with $|\psi'| = 1$. The $(3,2)$ dimensional supersphere $S^{(3,2)}$ can then be defined as

$$S^{(3,2)} \equiv \left\langle \psi' = \frac{\psi}{|\psi|} \in \mathcal{C}^{2,1} \setminus \{0\} \right\rangle. \tag{9.40}$$

Obviously $S^{(3,2)}$ has the 3-sphere S^3 as its even part.

The noncommutative $S^{(3,2)}$ is obtained by replacing ψ' by $\Psi \frac{1}{\sqrt{\widehat{N}}}$ where $\widehat{N} = a_i^\dagger a_i + b^\dagger b$ is the number operator. We have

$$\psi_\mu' \quad \longrightarrow \quad S_\mu := \Psi_\mu \frac{1}{\sqrt{\widehat{N}}} = \frac{1}{\sqrt{\widehat{N}+1}} \Psi_\mu \,,$$

$$\psi_\mu'^\dagger \quad \longrightarrow \quad S_\mu^\dagger := \frac{1}{\sqrt{\widehat{N}}} \Psi_\mu^\dagger = \Psi_\mu^\dagger \frac{1}{\sqrt{\widehat{N}+1}} \,, \tag{9.41}$$

where $\widehat{N} \neq 0$. Furthermore, we have that $[S_\mu, S_\nu\} = [S_\mu^\dagger, S_\nu^\dagger\} = 0$, while after a small calculation, we get

$$[S_\mu, S_\nu^\dagger\} = \frac{1}{\widehat{N}+1} \left(\delta_{\mu\nu} - (-1)^{|S_\mu||S_\nu|} S_\nu^\dagger S_\mu \right). \tag{9.42}$$

We note that as the eigenvalue of $\widehat{N}$ approaches to infinity, we recover $S^{(3,2)}$ back.

Noncommutative $S^{(3,2)}$ suffers from the same problem as noncommutative S^3 does: S_μ an $S_\mu^\dagger$ act on an infinite-dimensional Hilbert space so that we do not obtain finite-dimensional models for noncommutative $S^{(3,2)}$ either. Nevertheless, the structure of the non-commutative $S^{(3,2)}$ described above is quite useful in the construction of $S_F^{(2,2)}$ as well as for obtaining $*$-products on the "sections of bundles" over $S_F^{(2,2)}$ as we will discuss later in this chapter.

9.3.3 The Commutative Supersphere $S^{(2,2)}$

There is a supersymmetric generalization of the Hopf fibration. In this subsection we construct this (super)-Hopf fibration through studying the actions of $OSp(2,1)$ and $OSp(2,2)$ on $S^{(3,2)}$. We also establish that the supersphere $S^{(2,2)}$ is the adjoint orbit of $OSp(2,1)$, while it is a closely related (but not the adjoint) orbit of $OSp(2,2)$. We elaborate on the subtle features of the latter, which are important for future developments in this chapter.

We first note that the group manifold of $OSp(2,1)$ is nothing but $S^{(3,2)}$. Also note that $|\psi|^2$ is preserved under the group action $\psi \longrightarrow g\psi$ for $g \in OSp(2,1)$. Let us then consider the following map Π from the functions on the $(3,2)$-dimensional supersphere $S^{(3,2)}$ to functions on $S^{(2,2)}$:

$$\Pi \; : \; \psi' \longrightarrow \quad w_a(\psi,\bar\psi) := \bar\psi' \Lambda_a^{(\frac{1}{2})} \psi' = \frac{2}{|\psi|^2} \bar\psi \Lambda_a^{(\frac{1}{2})} \psi . \tag{9.43}$$

The fibres in this map are $U(1)$ as the overall phase in $\psi \to \psi e^{i\gamma}$ cancels out while no other degree of freedom is lost on r.h.s. Quotienting $S^{(3,2)} \equiv OSp(2,1)$ by the $U(1)$, fibres, we get the $(2,2)$ dimensional base space [†]

$$S^{(2,2)} := S^{(3,2)}/U(1) \equiv \left\{ w(\psi) = \big(w_1(\psi), \cdots, w_5(\psi)\big) \right\}. \tag{9.44}$$

Π is thus the projection map of the "super-Hopf fibration" over $S^{(2,2)}$ [117, 118, 78], and $S^{(2,2)}$ can be thought as the supersphere generalizing S^2.

We now characterize $S^{(2,2)}$ as an adjoint orbit of $OSp(2,1)$. First observe that $w(\psi)$ is a (super)-vector in the adjoint representation of $OSp(2,1)$. Under the action

$$w \to gw, \quad (gw)(\psi) = w(g^{-1}\psi), \quad g \in OSp(2,1) , \tag{9.45}$$

it transforms by the adjoint representation $g \to Ad\,g$:

$$w_a(g^{-1}\psi) = w_b\,(\psi)\,(Ad\,g)_{ba} . \tag{9.46}$$

The generators of $osp(2,1)$ in the adjoint representation are $ad\Lambda_a$ where

$$(ad\,\Lambda_a)_{cb} = if_{abc} . \tag{9.47}$$

From this and the infinitesimal variations $\delta w(\psi) = \varepsilon_a\,ad\,\Lambda_a\,w(\psi)$ of $w(\psi)$ under the adjoint action, where ε_i's are even and ε_α's are odd Grassmann variables, we can verify that

$$\delta(w_i(\psi)^2 + C_{\alpha\beta}w_\alpha(\psi)w_\beta(\psi)) = 0 . \tag{9.48}$$

Hence, $S^{(2,2)}$ is an $OSp(2,1)$ orbit with the invariant

$$\frac{1}{2}(w_a(S^{-1})_{ab}w_b) = w_i(\psi)^2 + C_{\alpha\beta}w_\alpha(\psi)w_\beta(\psi) . \tag{9.49}$$

[†]In what follows we do not show the $\bar\psi$ dependence of w_a to abbreviate the notation a little bit.

The value of the invariant can of course be changed by scaling. Now the even components of $w_a(\psi)$ are real while its odd entries depend on both θ and $\bar{\theta}$:

$$w_i(\psi) = \frac{1}{|\psi|^2}\bar{z}\sigma_i z \,, \;\; w_4(\psi) = -\frac{1}{|\psi|^2}(\bar{z}_1\theta + z_2\bar{\theta}) \,, \;\; w_5(\psi) = \frac{1}{|\psi|^2}(-\bar{z}_2\theta + z_1\bar{\theta}) \,.$$
(9.50)

From (9.36) and (9.50), one deduces the reality conditions

$$w_i(\psi)^\ddagger = w_i(\psi) \qquad w_\alpha(\psi)^\ddagger = -C_{\alpha\beta}\, w_\beta(\psi) \,.$$
(9.51)

The $OSp(2,1)$ orbit is preserved under this operation as can be checked directly using (9.51) in (9.49). The reality condition (9.51) reduces the degrees of freedom in $w_\alpha(\psi)$ to two. The $(3,2)$ number of variables $w_a(\psi)$ are further reduced to $(2,2)$ on fixing the value of the invariant (9.49). As $(2,2)$ is the dimension of $S^{(2,2)}$, there remains no further invariant in this orbit. Thus

$$S^{(2,2)} =$$
$$\left\langle \eta \in \mathbb{R}^{(3,2)} \,\Big|\, \eta_i^2 + C_{\alpha\beta}\,\eta_\alpha^{(-)}\eta_\beta^{(-)} = 1 \,, (\eta_i)^\ddagger = \eta_i \,, (\eta_\alpha^{(-)})^\ddagger = -C_{\alpha\beta}\eta_\beta^{(+)} \right\rangle \,,$$
(9.52)

where we have chosen $\frac{1}{4}$ for the value of the invariant. It is important to note that the superspace $\mathbb{R}^{(3,2)}$ in (9.52) is defined as the algebra of polynomials in generators η_i and $\eta_\alpha^{(-)}$ satisfying the reality conditions $\eta_i^\ddagger = \eta_i \,, \eta^{(-)\ddagger} = -C_{\alpha\beta}\eta_\beta^{(+)}$. Thus $S^{(2,2)}$ is embedded in $\mathbb{R}^{(3,2)}$ as described by (9.52).

As $OSp(2,2)$ acts on ψ, that is on $S^{(3,2)}$, preserving the $U(1)$ fibres in the map $S^{(3,2)} \to S^{(2,2)}$, it has an action on the latter. It is not the adjoint action, but closely related to it, as we now explain.

The nature of the $OSp(2,2)$ action on $S^{(2,2)}$ has elements of subtlety. If $g \in OSp(2,2)$ and $\psi \in S^{(3,2)}$ then $g\,\psi \in S^{(3,2)}$ and hence $w(g\,\psi) \in S^{(2,2)}$:

$$w_i(g\,\psi)^2 + C_{\alpha\beta}\, w_\alpha(g\,\psi)\, w_\beta(g\,\psi) = 1 \,,$$
$$w_i(g\,\psi)^\ddagger = w_i(g\,\psi) \,, \quad w_\alpha^\ddagger(g\,\psi) = -C_{\alpha\beta}\, w_\beta(g\,\psi) \,.$$
(9.53)

But the expansion of $w_\alpha(g\,\psi)$ for infinitesimal g contains not only the odd Majorana spinors $\eta_\alpha^{(-)}$, but also the even ones $\eta_\alpha^{(+)}$, where $(\eta_\alpha^{(+)})^\ddagger = -\sum_{\beta=6,7}\tilde{C}_{\alpha\beta}\,\eta_\beta^{(+)}$ ($\alpha = 6,7$). We cannot thus think of the $OSp(2,2)$ action as the adjoint action on the adjoint space of $OSp(2,1)$. The reason of course is that the Lie superalgebra $osp(2,1)$ is not invariant under graded commutation with the generators $\Lambda_{6,7,8}$ of $osp(2,2)$.

Now consider the generalization of the map (9.43) to the $osp(2,2)$ Lie algebra,

$$\psi' \longrightarrow \mathcal{W}_a(\psi) := \bar\psi' \Lambda_a^{(\frac{1}{2})} \psi' = \frac{2}{|\psi|^2} \bar\psi \Lambda_a^{(\frac{1}{2})} \psi, \quad a = (1,\ldots,8), \quad (9.54)$$

where the $\bar\psi$ dependence of $\mathcal{W}_a$ has been suppressed for notational brevity. Just as for $OSp(2,1)$, we find,

$$\mathcal{W}_a(g^{-1}\psi) = \mathcal{W}_b(\psi)(Ad\,g)_{ba}, \quad a,b = 1,\ldots,8, \quad g \in OSp(2,2). \quad (9.55)$$

Thus this extended vector $\mathcal{W}(\psi) = (\mathcal{W}_1(\psi), \mathcal{W}_2(\psi), \ldots, \mathcal{W}_8(\psi))$ transforms as an adjoint (super)-vector of $osp(2,2)$ under $OSp(2,2)$ action. The formula given in (9.50) extends to this case when index a there also takes the values $(6,7,8)$. Explicitly we have

$$\mathcal{W}_6(\psi) = \frac{1}{|\psi|^2}(\bar z_1 \theta - z_2 \bar\theta), \quad \mathcal{W}_7(\psi) = \frac{1}{|\psi|^2}(\bar z_2 \theta + z_1 \bar\theta),$$

$$\mathcal{W}_8(\psi) = 2\frac{1}{|\psi|^2}(\bar z_i z_i + 2\bar\theta\theta) = 2\left(2 - \frac{1}{|\psi|^2}\bar z_i z_i\right). \quad (9.56)$$

The reality conditions for $\mathcal{W}_6(\psi), \mathcal{W}_7(\psi), \mathcal{W}_8(\psi)$ are

$$\mathcal{W}_8(\psi)^\ddagger = \mathcal{W}_8(\psi), \quad \mathcal{W}_\alpha(\psi)^\ddagger = -\sum_{\beta=6,7} \tilde C_{\alpha\beta} \mathcal{W}_\beta(\psi), \quad \alpha = 6,7, \quad (9.57)$$

showing that the new spinor $\mathcal{W}_\alpha(\psi)$, $(\alpha = 6,7)$ is an even Majorana spinor as previous remarks suggested.

As $\mathcal{W}(\psi)$ transforms as an adjoint vector under $OSp(2,2)$, the $OSp(2,2)$ Casimir function is a constant on this orbit:

$$\frac{1}{2}(\mathcal{W}_a(S^{-1})_{ab}\mathcal{W}_b) = \mathcal{W}_i^2(\psi) + \tilde C_{\alpha\beta}\,\mathcal{W}_\alpha(\psi)\mathcal{W}_\beta(\psi) - \frac{1}{4}\mathcal{W}_8^2(\psi) = \text{constant}. \quad (9.58)$$

But we saw that the sum of the first term, and the second term with $\alpha, \beta = 4,5$ only, is invariant under $OSp(2,1)$. Hence so are the remaining terms:

$$\sum_{\alpha,\beta=6,7} \tilde C_{\alpha\beta}\mathcal{W}_\alpha(\psi)\mathcal{W}_\beta(\psi) - \frac{1}{4}\mathcal{W}_8(\psi)^2 = \text{constant}. \quad (9.59)$$

Its value is -1 as can be calculated by setting $\psi = (1,0,0)$.

In fact, since the $OSp(2,1)$ orbit has the dimension of $S^{(3,2)}/U(1)$ and $\mathcal{W}_a(\psi) = \mathcal{W}_a(\psi\,e^{i\gamma})$ are functions of this orbit, we can completely express the latter in terms of $w(\psi)$. We find[‡]

$$\mathcal{W}_\alpha(\psi) = -w_\beta \left(\frac{\sigma \cdot w(\psi)}{r}\right)_{\beta,\alpha-2}$$

$$\mathcal{W}_8(\psi) = \frac{2}{r}(r^2 + C_{\alpha\beta}w_\alpha w_\beta), \quad r^2 = w_i w_i. \quad (9.60)$$

[‡]$\mathcal{W}_{6,7,8}$ become $-\mathcal{W}_{6,7,8}$ for $J_{osp(2,2)-}$.

9.3.4   The Fuzzy Supersphere $S_F^{(2,2)}$

We are now ready to construct the fuzzy supersphere $S_F^{(2,2)}$. We do so by replacing the coordinates w_a of $S^{(2,2)}$ by $\hat{w}_a$:

$$w_a \longrightarrow \hat{w}_a = S^\dagger \Lambda_a^{(\frac{1}{2})} S = \frac{1}{\sqrt{\widehat{N}}} \Psi^\dagger \Lambda_a^{(\frac{1}{2})} \Psi \frac{1}{\sqrt{\widehat{N}}} = \frac{1}{\widehat{N}} \Psi^\dagger \Lambda_a^{(\frac{1}{2})} \Psi . \qquad (9.61)$$

Obviously, we have $\hat{w}_a$ commuting with the number operator $\widehat{N}$:

$$[\hat{w}_a , \widehat{N}] = 0 . \qquad (9.62)$$

Consequently, we can confine $\hat{w}_a$ to the subspace $\tilde{\mathcal{H}}_n$ of the Fock space of dimension $(2n+1)$ spanned by the kets

$$|n_1 , n_2 , n_3\rangle \equiv \frac{(a_1^\dagger)^{n_1}}{\sqrt{n_1!}} \frac{(a_2^\dagger)^{n_2}}{\sqrt{n_2!}} (b^\dagger)^{n_3}|0\rangle , \quad n_1 + n_2 + n_3 = n , \qquad (9.63)$$

where n_3 takes on the values 0 and 1 only. The Hilbert space $\tilde{\mathcal{H}}_n$ splits into the even subspace $\tilde{\mathcal{H}}_n^e$ and the odd subspace $\tilde{\mathcal{H}}_n^o$ of dimensions $n+1$ and n, respectively.

Linear operators, and hence w_a, acting on $\tilde{\mathcal{H}}_n$ generate the algebra of supermatrices $Mat(n+1, n)$ of dimension $(2n+1)^2$ which is customarily identified with the fuzzy supersphere. Similar to the fuzzy sphere, $S_F^{(2,2)}$ is also finite-dimensional: $Mat(n+1, n)$ is an inner product space with the inner product

$$(m_1 , m_2) = Str\, m_1^\dagger m_2 , \qquad m_i \in Mat(n+1, n) , \qquad (9.64)$$

where the identity matrix is already normalized to have the unit norm in this form.

In order to be more explicit, we first note that the $osp(2, 1)$ (and hence $osp(2, 2)$) Lie superalgebras can be realized as a supersymmetric generalization of the Schwinger construction by

$$\lambda_a = \Psi^\dagger (\Lambda_a^{(\frac{1}{2})}) \Psi , \quad [\lambda_a , \lambda_b\} = if_{abc}\lambda_c . \qquad (9.65)$$

The vector states in (9.63) for $n = 1$ give the superspin $J = \frac{1}{2}$ representation of $osp(2, 1)$, while for generic n they correspond to the n-fold graded symmetric tensor product of $J = \frac{1}{2}$ superspins that span the superspin $J = \frac{n}{2}$ representation of $osp(2, 1)$. Therefore, on the Hilbert space $\tilde{\mathcal{H}}_n$, we have

$$\left(\lambda_i \lambda_i + C_{\alpha\beta}\lambda_\alpha \lambda_\beta\right)\tilde{\mathcal{H}}_n = \frac{n}{2}\left(\frac{n}{2} + \frac{1}{2}\right)\tilde{\mathcal{H}}_n . \qquad (9.66)$$

Using the relation

$$\hat{w}_a \tilde{\mathcal{H}}_n = \frac{2}{n} \lambda_a \tilde{\mathcal{H}}_n \,, \tag{9.67}$$

we obtain

$$[\hat{w}_a , \hat{w}_b\} \tilde{\mathcal{H}}_n = \frac{2}{n} i f_{abc} \hat{w}_c \tilde{\mathcal{H}}_n \tag{9.68}$$

$$\left(\hat{w}_i \hat{w}_i + C_{\alpha\beta} \hat{w}_\alpha \hat{w}_\beta \right) \tilde{\mathcal{H}}_n = \left(1 + \frac{1}{n} \right) \tilde{\mathcal{H}}_n \,. \tag{9.69}$$

The radius $\sqrt{\left(1 + \frac{1}{n}\right)}$ of $S_F^{(2,2)}$ goes to 1 as n tends to infinity. The graded commutative limit is recovered when $J \to \infty$: in that limit, $[\hat{w}_a , \hat{w}_b\} \to 0$.

The Schwinger construction above naturally extends to the generators of $osp(2,2)$ as well. In general we can write

$$\widehat{\mathcal{W}}_a := \frac{2}{n} \lambda_a \,, \quad a = (1, \cdots, 8) \,. \tag{9.70}$$

(9.70) generate the $osp(2,2)$ algebra where

$$\widehat{\mathcal{W}}_a \to \mathcal{W}_a \quad \text{as} \quad n \to \infty \,. \tag{9.71}$$

The generators $\widehat{\mathcal{W}}_{6,7,8}$ can be realized in terms of the $osp(2,1)$ generators. This fact becomes important for field theories on both $S^{(2,2)}$ and $S_F^{(2,2)}$; Even though, these field theories have $OSp(2,1)$ invariance, the $osp(2,2)$ structure is needed to treat them as we will see later in the chapter.

The observables of $S_F^{(2,2)}$ are defined as the linear operators $\alpha \in Mat(n+1,n)$ acting on $Mat(n+1,n)$. They have the graded right- and left- action on the Hilbert space $Mat(n+1,n)$ given by

$$\alpha^L m = \alpha m \,, \quad \alpha^R m = (-1)^{|\alpha||m|} m\alpha \,, \quad \forall m \in Mat(n+1,n) \,. \tag{9.72}$$

They satisfy

$$(\alpha\beta)^L = \alpha^L \beta^L \,, \quad (\alpha\beta)^R = (-1)^{|\alpha||\beta|} \beta\alpha \,, \tag{9.73}$$

and commute in the graded sense:

$$[\alpha^L , \beta^R\} = 0 \,, \quad \forall \alpha , \beta \in Mat(n+1,n) \,. \tag{9.74}$$

In particular $osp(2,1)$ and $osp(2,2)$ act on $Mat(n+1,n)$ by the (super)-adjoint action:

$$ad\,\Lambda_a\, m = \left(\Lambda_a^L - \Lambda_a^R \right) m = [\Lambda_a , m\} \,, \tag{9.75}$$

which is a graded derivation on the algebra $Mat(n+1,n)$.

Before closing this section we note that left- and right-action of Ψ_μ and $\Psi_\mu^\dagger$ can also be defined on $Mat(n+1,n)$. They shift the dimension of the Hilbert space by 1 and will naturally arise in discussions of "fuzzy sections of bundles" in section 9.7.

9.4 More on Coherent States

In this section we construct the $OSp(2,1)$ supercoherent states (SCS) by projecting them from the coherent states associated to $\mathbb{C}^{2,1}$ [9]. In the literature the construction of $OSp(2,1)$ coherent states has been discussed [114, 115]. Here we explicitly show that our SCS is equivalent to the one obtained using the Perelomov's construction of the generalized coherent states, considered in chapter 3.

We start our discussion by introducing the coherent states including the bosonic and fermionic degrees of freedom [54, 53]:

$$|\psi\rangle \equiv |z, \theta\rangle = e^{-1/2|\psi|^2} e^{a_\alpha^\dagger z_\alpha + b^\dagger \theta} |0\rangle. \tag{9.76}$$

We can see from section 9.3 that the labels ψ of the states $|\psi\rangle$ are in one to one correspondence with points of the superspace $\mathcal{C}^{(2,1)}$. We recall that $|\psi|^2 \equiv |z_1|^2 + |z_2|^2 + \bar{\theta}\theta$. Hence $|\psi\rangle$'s are normalized to 1 as written.

The projection operator to the subspace $\tilde{\mathcal{H}}_n$ of the Fock space can be written as

$$P_n = \sum_{n=n_1+n_2+n_3} \frac{1}{n_1! n_2!} (a_1^\dagger)^{n_1} (a_2^\dagger)^{n_2} (b^\dagger)^{n_3} |0\rangle\langle 0| (b)^{n_3} (a_2)^{n_2} (a_1)^{n_1}, \tag{9.77}$$

where $n_3 = 0$ or 1. Clearly $P_n^2 = P_n$, $P_n^\dagger = P_n$.

Projecting $|\psi\rangle$ with P_n and renormalizing the result by the factor $(\langle\psi|P_n|\psi\rangle)^{-1/2}$, we get

$$|\psi', n\rangle = \frac{1}{\sqrt{n!}} \frac{(a_\alpha^\dagger z_\alpha + b^\dagger \theta)^n}{(|\psi|)^n} |0\rangle = \frac{(\Psi_\mu^\dagger \psi_\mu')^n}{\sqrt{n!}} |0\rangle. \tag{9.78}$$

This is the supercoherent state associated to $OSp(2,1)$. It is normalized to unity :

$$\langle\psi', n|\psi', n\rangle = 1. \tag{9.79}$$

We first establish the relation of (9.78) to Perelomov's construction of coherent states. To this end consider the following highest weight state in the $J_{osp(2,1)} = \frac{1}{2}$ representation of $osp(2,1)$ for which $\widehat{N} = 1$:

$$|J_{osp(2,1)} J_{su(2)}, J_3\rangle = |\frac{1}{2}, \frac{1}{2}, \frac{1}{2}\rangle. \tag{9.80}$$

This is also the highest weight state in the associated non-typical representation $J_{osp(2,2)+} = (\frac{1}{2})_{osp(2,2)+}$ of $osp(2,2)$. Consider now the action of the $OSp(2,1)$ on (9.80). This can be realized by taking $g \in OSp(2,1)$ and

$\mathcal{U}(g)$ as the corresponding element in the 3×3 fundamental representation. Thus let

$$|g\rangle = \mathcal{U}(g)|\frac{1}{2}, \frac{1}{2}, \frac{1}{2}\rangle, \qquad (9.81)$$

where $|g\rangle$ is the super-analogue of the Perelomov coherent state [54]. We can write

$$|\frac{1}{2}, \frac{1}{2}, \frac{1}{2}\rangle = \Psi_1^\dagger |0\rangle \qquad (9.82)$$

where $\Psi^\dagger = \left(\Psi_1^\dagger, \Psi_2^\dagger, \Psi_0^\dagger\right) \equiv \left(a_1^\dagger, a_2^\dagger, b^\dagger\right)$ as given in (9.38). In the basis spanned by the $\{\Psi_\mu^\dagger |0\rangle\}$, $(\mu = 1, 2, 0)$ the matrix of $\mathcal{U}(g)$ can be expressed as [114]

$$\mathcal{D}(g) = \begin{pmatrix} z_1' & -\bar{z}_2' & -\theta' \\ z_2' & \bar{z}_1' & -\bar{\theta}' \\ \chi & -\bar{\chi} & \lambda \end{pmatrix}, \qquad \sum_i |z_i'|^2 + \bar{\theta}'\theta' = 1. \qquad (9.83)$$

Then

$$|g\rangle = \left(\mathcal{D}(g)\right)_{1\mu} \Psi_\mu^\dagger |0\rangle$$
$$= (a_\alpha^\dagger z_\alpha' + b^\dagger \theta')|0\rangle = \Psi_\mu^\dagger \psi_\mu' |0\rangle. \qquad (9.84)$$

Clearly (9.84) is exactly equal to $|\psi', 1\rangle$ in (9.78).

For the case of general n, we start from the highest weight state $|\frac{n}{2}, \frac{n}{2}, \frac{n}{2}\rangle$ in the n-fold graded symmetric tensor product $\otimes_G^n$ of the $J_{osp(2,1)} = \frac{1}{2}$ representation and the corresponding representative $\mathcal{U}^{\otimes_G^n}(g)$ of g:

$$|\frac{n}{2}, \frac{n}{2}, \frac{n}{2}\rangle := |\frac{1}{2}, \frac{1}{2}, \frac{1}{2}\rangle \otimes_G \cdots\cdots \otimes_G |\frac{1}{2}, \frac{1}{2}, \frac{1}{2}\rangle,$$
$$\mathcal{U}^{\otimes_G^n}(g) := \mathcal{U}(g) \otimes_G \cdots\cdots \otimes_G \mathcal{U}(g). \qquad (9.85)$$

Note that, since $\mathcal{U}(g)$ is an element of $OSp(2,1)$, it is even. The corresponding coherent state is

$$|g; \frac{n}{2}\rangle = \mathcal{U}^{\otimes_G^n}|\frac{n}{2}, \frac{n}{2}, \frac{n}{2}\rangle = \mathcal{U}(g)|\frac{1}{2}, \frac{1}{2}, \frac{1}{2}\rangle \otimes_G \cdots\cdots \otimes_G \mathcal{U}(g)|\frac{1}{2}, \frac{1}{2}, \frac{1}{2}\rangle. \qquad (9.86)$$

Upon using (9.84), this becomes equal to (9.78) as we intended to show.

The coherent state in (9.76) can be written as a sum of its even and odd components by expanding it in powers of $b^\dagger$:

$$|\psi\rangle \equiv |z, \theta\rangle = e^{-1/2|\psi|^2} e^{a_\alpha^\dagger z_\alpha} \left(|0,0\rangle - \theta|0,1\rangle\right)$$
$$= |z,0\rangle - \theta|z,1\rangle. \qquad (9.87)$$

We proved in chapter 2 that the diagonal matrix elements of an operator K in the coherent states $|z\rangle$ completely determine K. That proof can be adapted to $|\psi\rangle$ as can be infered from (9.87). It can next be adapted to $|\psi', n\rangle$ for operators leaving the the subspace $N = n$ invariant. The line of reasoning is similar to the one used for $SU(2)$ coherent states in chapter 2.

9.5   The Action on the Supersphere $S^{(2,2)}$

The simplest $Osp(2,1)$-invariant Lagrangian density $\mathcal{L}$ can be written as $\Phi^{\ddagger}V\Phi$, where Φ is the scalar superfield and V an appropriate differential operator. We focus on $\mathcal{L}$ in what follows.

The superfield Φ is a function on $S^{(2,2)}$, that is , it is a function of w_a , $(a = 1, 2, \cdots, 5)$ fulfilling the constraint in (9.51).

For functional integrals, what is important is not $\mathcal{L}$, but the action S. Thus we need a method to integrate $\mathcal{L}$ over $S^{(2,2)}$ maintaining SUSY.

We also need a choice of V to find S. The appropriate choice is not obvious, and was discovered by Fronsdal [119]. It was adapted to $Osp(2,1)$ by Grosse et al. [7].

We now describe these two aspects of S and indicate also the calculation of S.

i. Integration on $S^{(2,2)}$

Let K be a scalar superfield on $S^{(2,2)}$. It is a function of w_i and w_α. We can write it as

$$K = k_0 + C_{\alpha\beta}k_\alpha w_\beta + k_1 C_{\alpha\beta}w_\alpha w_\beta \qquad (9.88)$$

where k_0 and k_1 are even, $k_\alpha(\alpha = 4, 5)$ are odd and $k's$ do not depend on w_α's, but can depend on w_i's.

There is no need to include $w_{6,7}$ in (9.88) as they are nonlinearly related to $w_{4,5}$.

The integral of K over $S^{(2,2)}$ (of radius R) can be defined as

$$I(K) = \int d\Omega \, r^2 \, dr \, dw_4 \, dw_5 \, \delta(r^2 + C_{\alpha\beta}w_\alpha w_\beta - R^2) \, K \qquad (9.89)$$

where $R > 0$ and $d\Omega = d\cos(\theta)d\psi$ is the volume form on S^2.

In the volume form in the integrand of $I(K)$, we do not constrain w_i , w_α to fulfil $w_i^2 + C_{\alpha\beta}w_\alpha w_\beta = R^2$. This constraint is taken care of by the δ-function.

The grade-adjoint representation of $osp(2,1)$ is 5-dimensional. It acts on $\mathbb{R}^{3,2} := \mathbb{R}^3 \oplus \mathbb{R}^2$ with an even subspace $\mathbb{R}^3$ (spanned by w_i) and an odd subspace $\mathbb{R}^2$ (spanned by w_α). Integration in (9.89) uses the $OSp(2,1)$-invariant volume form on $\mathbb{R}^{3,2}$ and the $OSp(2,1)$-invariant δ-function to restrict the integral to $S^{(2,2)}$. Thus $I(K)$ is invariant under the action of SUSY on K.

$I(K)$ is in fact $OSp(2,2)$ invariant. That is because $OSp(2,2)$ leaves the argument of the δ-function invariant as we already saw. The volume form as well is invariant because of the nonlinear realization of $w_{6,7,8}$ as is easily checked.

We can write

$$\delta(r^2 + C_{\alpha\beta}w_\alpha w_\beta - R^2) = \delta(r^2 - R^2) + 2w_4 w_5 \frac{d}{dr^2}\delta(r^2 - R^2)$$

$$= \frac{1}{2R}\delta(r - R) + \frac{1}{2Rr}w_4 w_5 \frac{d}{dr}\delta(r - R) \tag{9.90}$$

where we have dropped terms involving $\delta(r + R)$ and $\frac{d}{dr}\delta(r + R)$ as they do not contribute to the dr-integral from 0 to ∞. Thus using also

$$\int dw_4 \, dw_5 \, w_4 \, w_5 = -1\,, \tag{9.91}$$

we get

$$I(K) = \int d\Omega \left[\frac{d}{dr}(rk_0) - Rk_1\right]_{r=R}. \tag{9.92}$$

This is a basic formula.

ii. The $OSp(2,1)$-invariant operator V

The first guess would be the Casimir K_2 of $OSp(2,1)$, written in terms of differential and superdifferential operators [119, 7]. But this choice is not satisfactory. The simplest $OSp(2,1)$-invariant action is that of the Wess-Zumino model [120] and contains just the standard quadratic ("kinetic energy") terms of the scalar and spinor fields. But K_2 gives a different action with nonstandard spinor field terms [119, 7].

But any $OSp(2,1)$ representation is also a nontypical representation of $OSp(2,2)$ and its $OSp(2,2)$ Casimir K_2' is certainly $OSp(2,1)$ invariant. Thus so is

$$V := K_2' - K_2 = \Lambda_6\Lambda_7 - \Lambda_7\Lambda_6 + \frac{1}{4}\Lambda_8^2\,. \tag{9.93}$$

It happens that this V correctly reproduces the needed simple action.

iii. How to calculate : A sketch

SUSY calculations are typically a bit tedious. For that reason, we just sketch the details and give the final answer.

We first expand the superfield Φ in the standard manner:

$$\Phi(w_i, w_\alpha) = \varphi_0(w_i) + C_{\alpha\beta}\psi_\alpha w_\beta + \chi(w_i)C_{\alpha\beta}w_\alpha w_\beta\,. \tag{9.94}$$

Here $(\alpha, \beta = 4, 5)$, φ_0 and χ are even fields (commuting with w_α) and ψ_α are odd fields (anticommuting with w_α).

The aim is to calculate

$$S = I(\Phi^\ddagger V \Phi) \,. \tag{9.95}$$

For V we take (9.93) where $\Lambda_{6,7,8}$ represent the $OSp(2,2)$ generators acting on w_i, w_α. Thus we need to know how they act on the constituents of Φ in (9.94).

The action of Λ_α on w_β follows from (9.16) since w_β transform like $osp(2,2)$ generators:

$$\Lambda_\alpha w_i = \frac{1}{2} w_\beta (\tilde{\sigma}_i)_{\beta\alpha} \,, \quad \Lambda_\alpha w_{\beta-2} = \frac{1}{2} C_{\alpha\beta} w_8 \,, \quad \alpha, \beta = 6, 7 \,. \tag{9.96}$$

We now write $w_{6,7}$ in terms of $w_{4,5}$ using the relation (9.60) to find

$$\Lambda_\alpha w_i = -\frac{1}{2} w_{\gamma-2} (\sigma \cdot \hat{w})_{\gamma\beta} (\tilde{\sigma}_i)_{\beta\alpha} \,,$$

$$\Lambda_\alpha w_{\beta-2} = \frac{1}{2} C_{\alpha\beta} \frac{2}{r} (r^2 + 2 w_4 w_5) \,, \quad \alpha, \beta, \gamma = 6, 7 \,. \tag{9.97}$$

The action of of Λ_α on the fields of (9.94) follows from the chain rule. For example,

$$\Lambda_\alpha \varphi_0(w_i) = (\Lambda_\alpha w_i) \frac{\partial}{\partial w_i} \varphi_0(w_i) \,. \tag{9.98}$$

The ingredients for working out the action are now at hand. The calculation can be conveniently done for a real superfield:

$$\Phi^\ddagger = \Phi \,. \tag{9.99}$$

Φ can be decomposed in component fields as follows:

$$\Phi = \psi_0 + C_{\alpha\beta} \psi_\alpha \theta_\beta + \frac{1}{2} \chi C_{\alpha\beta} \theta_{\alpha\beta} \tag{9.100}$$

Then with θ_α an odd Majorana spinor,

$$\theta_\alpha^\ddagger = -C_{\alpha\beta} \theta_\beta \,, \tag{9.101}$$

we find that so is ψ:

$$\psi_\alpha^\ddagger = -C_{\alpha\beta} \psi_\beta \,. \tag{9.102}$$

We give the answer for the action

$$S(\Phi) = \int d\Omega \, r^2 \, dr \, \delta(r^2 + C_{\alpha\beta} w_\alpha w_\beta - 1) \Phi V \Phi \,. \tag{9.103}$$

We have set $R = 1$ whereas in previous sections we had $R = \frac{1}{2}$. We have

$$S(\Phi) = \int d\Omega \left\{ -\frac{1}{4} (\mathcal{L}\varphi_0)^2 + \frac{1}{4} (\chi - \varphi_0')^2 - \frac{1}{4} (C\psi)_\alpha (D\psi)_\alpha \right\}$$

$$\varphi_0' = \frac{1}{r} \frac{d}{dr} \psi_0 \,, \quad D = -\tilde{\sigma} \cdot \mathcal{L} + 1 \,, \quad \mathcal{L}_i = i(\vec{r} \times \vec{\nabla})_i \,. \tag{9.104}$$

The Dirac operator D here is unitarily equivalent to the Dirac operator in chapter 8.

$(\chi_0 - \varphi_0')$ is the auxiliary field F. Having no kinetic energy term, it can be eliminated. SUSY transformations mix all the fields.

A complex superfield Φ can be decomposed into two real superfields:

$$\Phi = \Phi^{(1)} + i\Phi^{(2)} \tag{9.105}$$

$$\Phi = \frac{\Phi + \Phi^{\ddagger}}{2} \quad , \Phi^{(2)} = \frac{\Phi - \Phi^{\ddagger}}{2i} \tag{9.106}$$

The action for Φ is the sum of the actions for $\Phi^{(i)}$. We can use (9.104) to write it. No separate calculation is needed.

9.6 The Action on the Fuzzy Supersphere $S_F^{(2,2)}$

Finding the action on $S_F^{(2,2)}$ is the crucial step for regularizing supersymmetric field theories using finite-dimensional matrix models, preserving $OSp(2,1)$-invariance.

We have seen that S^2 and S_F^2 allow instanton sectors. They affect chiral symmetry and are important for physics.

There are SUSY generalizations of these instantons. They are discussed in Chapter 10.

9.6.1 *The Integral and Supertrace*

In fuzzy physics with no SUSY, trace substitutes for $SU(2)$-invariant integration. The trace trM of an $(n+1) \times (n+1)$ matrix M is invariant under the $SU(2)$ action $M \to U(g)MU(g)^{-1}$ by its angular momentum $\frac{n}{2}$ representation $SU(2) : g \to U(g)$. It becomes the invariant integration in the large n-limit.

In fuzzy SUSY physics, the corresponding $OSp(2,2)$ invariant trace is supertrace str.

But (9.104) gives invariant integration in the (graded) commutative limit. We now establish that str goes over to the invariant integration as the cut-off $n \to \infty$.

A simple way to establish this is to use the supercoherent states. We have already defined them in (9.78). Here we drop the $\prime$ on ψ and write

$$|\psi, n\rangle = \frac{(a_\alpha^\dagger z_\alpha + b^\dagger \theta)^n}{\sqrt{n!}} |0\rangle . \tag{9.107}$$

This ψ is to be distinguished from those in (9.100).

Then as we saw, to every operator $\hat{K}$ commuting with $N = a_i^\dagger a_i + b^\dagger b$, we can define its symbol K, a function of $w's$, by

$$K(w) = \langle \psi, N | \hat{K} | \psi, N \rangle. \tag{9.108}$$

An invariant "integral" $\hat{I}$ on $\hat{K}$ can then be defined as

$$\hat{I}(\hat{K}) = I(K). \tag{9.109}$$

With the normalization

$$\int d\Omega = 1 \quad \text{or} \quad d\Omega = \frac{d\cos\theta \wedge d\phi}{4\pi}, \tag{9.110}$$

we can show that

$$\hat{I}(\hat{K}) = \frac{1}{2} str K. \tag{9.111}$$

It is then clear that str becomes $2I$ as $n \to \infty$.

The proof is easy. First note that for the non-SUSY coherent state

$$|z, n\rangle = \frac{(a^\dagger \cdot \hat{z}^n)}{\sqrt{n!}} |0\rangle, \quad \hat{z} \cdot \hat{z} = 1. \tag{9.112}$$

$$\int d\Omega \langle \hat{z}, n | \hat{A} | \hat{z}, n \rangle = \frac{1}{n+1} Tr \hat{A} \tag{9.113}$$

if $\hat{A}$ is an operator on the subspace spanned by $|\hat{z}, n\rangle$ for fixed n.

Terms linear in b and $b^\dagger$ have zero str. Hence we can assume that

$$\hat{K} = M_0 + M_1 b^\dagger b \tag{9.114}$$

where M_j are polynomials in $a_i^\dagger a_j$.

It can be easily checked that $str\hat{K}$ is $OSp(2,2)$-invariant as well.

In the $OSp(2,1)$ IRR $\left[\frac{N}{2}\right]_{osp(2,1)}$, the even subspace of its carrier space has angular momentum $\frac{N}{2}$ and odd subspace has angular momentum $\frac{N-1}{2}$. Hence

$$str\hat{K} = tr_{N+1} M_0 - tr_N M_0 - tr_N M_1 \tag{9.115}$$

where tr_m indicates trace over an m-dimensional space.

As for $\hat{I}(\hat{K})$, we note that

$$|\psi, N\rangle = |z, N\rangle + \sqrt{N} b^\dagger \theta | z, N - 1\rangle. \tag{9.116}$$

Hence

$$K(w) = \langle z, N | M_0 | z, N\rangle + N \bar{\theta}\theta \langle z, N - 1 | M_0 | z, N - 1\rangle$$
$$+ N \bar{\theta}\theta \langle z, N - 1 | M_1 | z, N - 1\rangle. \tag{9.117}$$

But by (9.50), $\bar{\theta}\theta = w_4 w_5$. So on using (9.92), we get

$$I(K) = -\frac{1}{2} \int d\Omega \Big\{ N \langle z, N - 1 | M_0 | z, N - 1\rangle + N \langle z, N - 1 | M_1 | z, N - 1\rangle$$
$$-(N+1)\rangle z, N | M_0 | z, N \Big\} = \frac{1}{2} str \hat{K} \tag{9.118}$$

as claimed.

9.6.2 *OSp(2, 1) IRR's with Cut-Off N*

The Clebsch-Gordan series for $OSp(2,1)$ is

$$[J]_{osp(2,1)} \otimes [K]_{osp(2,1)} =$$
$$[J + K]_{osp(2,1)} \oplus \left[J + K - \frac{1}{2}\right]_{osp(2,1)} \oplus \cdots \oplus [|J - K|]_{osp(2,1)} \quad (9.119)$$

The series on R.H.S thus descends in steps of $\frac{1}{2}$ (and not in steps of 1 as for $su(2)$) from $J + K$ to $|J - K|$.

Under the (graded) adjoint action of $osp(2,1)$, the linear operators in the representation space of $\left[\frac{N+1}{2}\right]_{osp(2,1)}$ transform as $\left[\frac{N+1}{2}\right]_{osp(2,1)} \otimes \left[\frac{N+1}{2}\right]_{osp(2,1)}$. Hence the $osp(2,1)$ content of the fuzzy supersphere is

$$\left[\frac{N+1}{2}\right]_{osp(2,1)} \otimes \left[\frac{N+1}{2}\right]_{osp(2,1)} =$$
$$[N + 1]_{osp(2,1)} \oplus \left[\frac{N+1}{2}\right]_{osp(2,1)} \oplus \left[N + \frac{1}{2}\right]_{osp(2,1)} \oplus \cdots \oplus [0]_{osp(2,1)}.$$
$$(9.120)$$

We now discuss

- The highest weight angular momentum states in each of these IRR's and the realization of $osp(2,2)$ on these $osp(2,1)$ multiplets, and
- The spectrum of V and the free supersymmetric scalar field action on the fuzzy supersphere.

9.6.3 *The Highest Weight States and the osp(2,2)-Invariant Action*

The graded Lie algebra $osp(2,1)$ is of rank 1. We can diagonalize (a multiple of) one operator in $osp(2,1)$ in each IRR. We choose it to be Λ_3, the third component of angular momentum.

Λ_4 is a raising operator for Λ_3, raising its eigenvalues by $\frac{1}{2}$. The vector state annihilated by Λ_4 in an IRR of $osp(2,1)$ is it highest weight state.

$\Lambda_+ = \Lambda_1 + i\Lambda_2$ is also a raising operator for Λ_3, raising its eigenvalue by $+1$. Vector states annihilated by Λ_4 are the highest weight states for the $su(2)$ IRR's contained in an $osp(2,1)$ IRR. A vector state in an IRR annihilated by Λ_4 is also annihilated by Λ_+.

The matrices of the fuzzy supersphere are polynomials in $a_i^\dagger a_j$, $a_i^\dagger b$, $b^\dagger a_i$ restricted to the subspace with $N = a_i^\dagger a_i + b^\dagger b$ fixed. Supersymmetry acts on them by adjoint action. The expression for Λ_4 is given in (9.65) while

$$\Lambda_+ = a_1^\dagger a_2 . \tag{9.121}$$

It follows that for J integral,

The highest weight state for $[J]_{osp(2,1)} = (a_1^\dagger a_2)^J$,

the highest weight state for $[J - \frac{1}{2}]_{osp(2,1)} = (a_1^\dagger a_2)^{J-1}\Lambda_6$.   (9.122)

The fact that $(a_1^\dagger a_2)^{J-1}\Lambda_6$ anticommutes with Λ_4 follows from $\{\Lambda_4 , \Lambda_6\} = 0$.

The states with angular momentum $J - \frac{1}{2}$ in $[J]_{osp(2,1)}$ and $J - 1$ in $[J - \frac{1}{2}]_{osp(2,1)}$ which are $su(2)$-highest weight states can be got acting with $ad\Lambda_5$ on highest weight states in (9.122). Thus

$$[J]_{osp(2,1)} :   (a_1^\dagger a_2)^J \quad \xrightarrow{ad\Lambda_5} \quad (a_1^\dagger a_2)^{J-1}\Lambda_4$$

$$ad\Lambda_7 \downarrow \quad \nearrow ad\Lambda_8 \quad ad\Lambda_7 \downarrow \tag{9.123}$$

$$[J - \tfrac{1}{2}]_{osp(2,1)} : (a_1^\dagger a_2)^{J-1}\Lambda_6 \quad \xrightarrow{ad\Lambda_5} \quad X$$

where

$$X = \frac{1 + N - J}{4}(a_1^\dagger a_2)^{J-1} + \frac{2J - 1}{4}(a_1^\dagger a_2)^{J-1}b^\dagger b . \tag{9.124}$$

As usual, ad denotes graded adjoint action as in 9.75. The vectors are not normalized. The arrows indicate the adjoint actions of $\Lambda_{5,7,8}$. They establish that $osp(2,2)$ acts irreducibly on $[J]_{osp(2,1)} \oplus [\frac{J-1}{2}]_{osp(2,1)}$.

We also see that

$$J = \left(0, \frac{1}{2}, \cdots , \frac{N+1}{2}\right) . \tag{9.125}$$

9.6.4   *The Spectrum of V*

We show that for J integer

$$V\big|_{[J]_{osp(2,1)}} = \frac{J}{2}\mathbf{1} , \tag{9.126a}$$

$$V\big|_{[J-\frac{1}{2}]_{osp(2,1)}} = -\frac{J}{2}\mathbf{1} . \tag{9.126b}$$

Proof of (9.126a)

It is enough to evaluate V on the highest weight state $(a_1^\dagger a_2)^J$. Since

$$ad\Lambda_8 \, (a_1^\dagger a_2)^J = ad\Lambda_6 \, (a_1^\dagger a_2)^J = 0 \,, \tag{9.127}$$

we have

$$\begin{aligned} V(a_1^\dagger a_2)^J &= (ad\Lambda_6 ad\Lambda_7 - ad\Lambda_7 ad\Lambda_6)(a_1^\dagger a_2)^J \\ &= (ad\Lambda_6 ad\Lambda_7 + ad\Lambda_7 ad\Lambda_6)(a_1^\dagger a_2)^J \\ &= ad\{\Lambda_6 \,, \Lambda_7\} \, (a_1^\dagger a_2)^J \\ &= -\frac{1}{2}(\varepsilon\sigma_i)_{67} \, ad\Lambda_i (a_1^\dagger a_2)^J \\ &= \frac{1}{2} ad\Lambda_3 \, (a_1^\dagger a_2)^J \\ &= \frac{J}{2}(a_1^\dagger a_2)^J \,. \end{aligned} \tag{9.128}$$

Proof of (9.126b)

We evaluate V on $(a_1^\dagger a_2)^{J-1}\Lambda_6$. We have

$$ad\Lambda_8 \, (a_1^\dagger a_2)^{J-1}\Lambda_6 = (a_1^\dagger a_2)^{J-1}\Lambda_4 \,. \tag{9.129}$$

Thus

$$\frac{1}{4}(ad\Lambda_8)^2 (a_1^\dagger a_2)^{J-1}\Lambda_6 = \frac{1}{4}(a_1^\dagger a_2)^{J-1}\Lambda_6 \,. \tag{9.130}$$

Now the $osp(2,1)$ Casimir K_2 has value $J(J+\frac{1}{2})\mathbf{1}$ in the IRR $[J]_{osp(2,1)}$ while

$$(ad\Lambda_i)^2 (a_1^\dagger a_2)^{J-1}\Lambda_4 = (J-\frac{1}{2})(J+\frac{1}{2}) \, (a_1^\dagger a_2)^{J-1}\Lambda_4 \,. \tag{9.131}$$

Hence with $\alpha \,, \beta \in [4,5]$,

$$(\varepsilon_{\alpha\beta} \, ad\Lambda_\alpha \, ad\Lambda_\beta) \, (a_1^\dagger a_2)^{J-1}\Lambda_4 = \frac{2J+1}{4}(a_1^\dagger a_2)^{J-1}\Lambda_4 \,. \tag{9.132}$$

But

$$e^{i\frac{\pi}{2}\Lambda_8} \, \Lambda_{4,5} \, e^{-i\frac{\pi}{2}\Lambda_8} = i \, \Lambda_{6,7} \,. \tag{9.133}$$

Hence

$$e^{i\frac{\pi}{2}\Lambda_8} \, (\varepsilon_{\alpha\beta} \, ad\Lambda_\alpha \, ad\Lambda_\beta \, (a_1^\dagger a_2)^{J-1}\Lambda_4) \, e^{-i\frac{\pi}{2}\Lambda_8} =$$

$$-(\varepsilon_{\alpha\beta} \, ad\Lambda_{\alpha'} \, ad\Lambda_{\beta'}) \, (i(a_1^\dagger a_2)^{J-1}\Lambda_6) = \frac{2J+1}{4} \, i \, (a_1^\dagger a_2)^{J-1}\Lambda_6 \tag{9.134}$$

(9.126b) follows upon using this and (9.131).

9.6.5   *The Fuzzy SUSY Action*

Let J be integral. We can write the highest weight component in angular momentum J of the superfield in the IRR $[J]_{osp(2,1)}$ as

$$\Phi_J = c_j(a_1^\dagger a_2)^J + (a_1^\dagger a_2)^{J-1}\xi_{J-\frac{1}{2}}\Lambda_4 \qquad (9.135)$$

where c_j is a (commuting) complex number and $\xi_{J-\frac{1}{2}}$ is a Grassmann number. The $osp(2,2)$ transformations map $[J]_{osp(2,1)}$ to $[J-\frac{1}{2}]_{osp(2,1)}$. The highest weight component in the latter can be written as

$$\Phi_{J-\frac{1}{2}} = \eta_{J-\frac{1}{2}}(a_1^\dagger a_2)^{J-1}\Lambda_5 + d_{J-1}X\,, \qquad (9.136)$$

where $\eta_{J-\frac{1}{2}}$ is a Grassmann and d_{J-1} a complex number.

The fuzzy action for the highest weight state in $[J]_{osp(2,1)}$ is

$$
\begin{aligned}
S_F^J(M=J) &= \frac{J}{2}str\Phi_J^\dagger\Phi_J \\
&= \frac{J}{2}\Big[\xi_{J-\frac{1}{2}}^\dagger\xi_{J-\frac{1}{2}}\,str[\Lambda_4^\dagger\,(a_2^\dagger a_1)^{J-1}\,(a_1^\dagger a_2)^{J-1}] \\
&\quad + |c_J|^2 str[(a_2^\dagger a_1)^J(a_1^\dagger a_2)^J]\Big]\,, \quad \Lambda_4^\dagger = -\Lambda_5\,, \quad (9.137)
\end{aligned}
$$

since the two terms in Φ_J are str-orthogonal. For $[J-\frac{1}{2}]_{osp(2,1)}$, instead,

$$
\begin{aligned}
S_F^{J-\frac{1}{2}}(M=J-\frac{1}{2}) &= \frac{J}{2}str\Phi_{J-\frac{1}{2}}^\dagger\Phi_{J-\frac{1}{2}} \\
&= -\frac{J}{2}\Big[\eta_{J-\frac{1}{2}}^\dagger\eta_{J-\frac{1}{2}}\,str[\Lambda_5^\dagger(a_2^\dagger a_1)^{J-1}(a_1^\dagger a_2)^{J-1}\Lambda_5] \\
&\quad + |d_{J-1}|^2\,strX^\dagger X\Big]\,, \quad \Lambda_5^\dagger = \Lambda_4\,, \quad (9.138)
\end{aligned}
$$

since the two terms in $\Phi_{J-\frac{1}{2}}$ are also str-orthogonal. The second term here is the integral spin term. It is positive since

$$str\,X^\dagger X < 0 \qquad (9.139)$$

as can be verified.

Str-orthogonality extends also to Φ_J and $\Phi_{J-\frac{1}{2}}$:

$$str\Phi^\dagger\Phi_{J-\frac{1}{2}} = 0\,. \qquad (9.140)$$

Hence for the highest weight states of $[J]_{osp(2,2)} = [J]_{osp(2,1)} \oplus [J-\frac{1}{2}]_{osp(2,1)}$, the actions add up:

$$S_F^{J\oplus(J-\frac{1}{2})}\ \text{for highest weight states} \ = S_F^J(J) + S_F^{J-\frac{1}{2}}(J-\frac{1}{2})\,. \quad (9.141)$$

The superfield Φ is a superposition of such terms. We must first include all angular momentum descendants of Φ_J and $\Phi_{J-\frac{1}{2}}$. We must also sum on J from 0 to N in steps of $\frac{1}{2}$.

For the fuzzy sphere S_F^2, such calculations are best performed using spherical tensors $\widehat{T}_{LM}(N)$ and their properties. Similarly, perhaps such calculations are best performed on the fuzzy supersphere using supersymmetric spherical tensors. But as yet only certain basic results about these tensors are available [8].

Reality conditions like $\Phi^{\ddagger} = \Phi$ constrain the Fourier coefficients $c_j, \xi_{J-\frac{1}{2}}, \eta_{J-\frac{1}{2}}, d_{J-1}$.

9.7 The $*$-Products

9.7.1 *The $*$-Product on $S_F^{(2,2)}$*

The diagonal matrix elements of operators in the supercoherent state $|\psi', n\rangle$ define functions on $S_F^{(2,2)}$. The $*$-product of functions on $S_F^{(2,2)}$ is induced by this map of operators to functions. To determine this map explicitly, it is sufficient to compute the matrix elements of the operators $\widehat{\mathcal{W}}_a$. Generalization to arbitrary operators can then be made easily as we will see.

The diagonal coherent state matrix element for $\widehat{\mathcal{W}}_a$'s are

$$\mathcal{W}_a(\psi', \bar{\psi}', n) = \langle \psi', n | \widehat{\mathcal{W}}_a | \psi', n \rangle = \frac{2}{|\psi|^2} \bar{\psi} \Lambda_a^{(\frac{1}{2})} \psi = \bar{\psi}' \Lambda_a^{(\frac{1}{2})} \psi'. \quad (9.142)$$

This defines the map

$$\widehat{\mathcal{W}}_a \longrightarrow \mathcal{W}_a \quad (9.143)$$

of the operator $\widehat{\mathcal{W}}_a$ to functions $\mathcal{W}_a$. $\mathcal{W}_a$ is a superfunction on $S_F^{(2,2)}$ since it is invariant under the $U(1)$ phase $\psi' \to \psi' e^{i\gamma}$.

We are now ready to define and compute the $*$-product of two functions of the form $\mathcal{W}_a$ and $\mathcal{W}_b$. It depends on n, and to emphasise this we include it in the argument of the product. It is given by

$$\mathcal{W}_a * \mathcal{W}_b(\psi', \bar{\psi}', n) = \langle \psi', n | \widehat{\mathcal{W}}_a \widehat{\mathcal{W}}_b | \psi', n \rangle \quad (9.144)$$

which becomes, after a little manipulation,

$$\mathcal{W}_a *_n \mathcal{W}_b(\psi', \bar{\psi}', n) = \frac{1}{n} \bar{\psi}' \left(\Lambda_a^{(\frac{1}{2})} \Lambda_b^{(\frac{1}{2})} \right) \psi' + \frac{n-1}{n} \left(\bar{\psi}' \Lambda_a^{(\frac{1}{2})} \psi' \right) \left(\bar{\psi}' \Lambda_b^{(\frac{1}{2})} \psi' \right).$$
$$(9.145)$$

Furthermore, since $\psi' \Lambda_a^{(\frac{1}{2})} \Lambda_b^{(\frac{1}{2})} \psi'$ is $\mathcal{W}_a * \mathcal{W}_b(\psi', \bar{\psi}', 1)$, (9.145) can be rewritten as

$$\mathcal{W}_a *_n \mathcal{W}_b \left(\psi', \bar{\psi}', n\right) = \frac{1}{n} \mathcal{W}_a *_1 \mathcal{W}_b \left(\psi', \bar{\psi}', 1\right) + \frac{n-1}{n} \mathcal{W}_a \left(\psi', \bar{\psi}'\right) \mathcal{W}_b \left(\psi', \bar{\psi}'\right).$$
$$(9.146)$$

Introducing the matrix K with

$$K_{ab} := \mathcal{W}_a *_1 \mathcal{W}_b - \mathcal{W}_a \mathcal{W}_b, \qquad (9.147)$$

we can express (9.146) as

$$\mathcal{W}_a *_n \mathcal{W}_b = \frac{1}{n} K_{ab} + \mathcal{W}_a \mathcal{W}_b. \qquad (9.148)$$

In this form it is apparent that in the graded commutative limit $n \to \infty$, we recover the graded commutative product of functions $\mathcal{W}_a$ and $\mathcal{W}_b$.

The *-product of arbitrary functions on $S_F^{(2,2)}$ can be obtained via a similar procedure used to derive that on S_F^2. In this case, one also needs to pay attention to the graded structure of the operators. Thus we can start from the generic operators F and G in the representation $\left(\frac{n}{2}\right)_{osp(2,2)+}$ expressed as

$$\widehat{F} = F^{a_1 a_2 \cdots a_n} \widehat{\mathcal{W}_{a_1}} \otimes_G \cdots \otimes_G \widehat{\mathcal{W}_{a_n}},$$
$$\widehat{G} = G^{b_1 b_2 \cdots b_n} \widehat{\mathcal{W}_{b_1}} \otimes_G \cdots \otimes_G \widehat{\mathcal{W}_{b_n}}, \qquad (9.149)$$

where for example $F^{a_1 \cdots a_i a_j \cdots a_n} = (-1)^{|a_i||a_j|} F^{a_1 \cdots a_j a_i \cdots a_n}$, $|a_i| \, (mod\, 2)$ being the degree of the index a_i. After a long but a straightforward calculation, the following finite-series formula is obtained (details can be found in [9]):

$$\mathcal{F}_n *_n \mathcal{G}_n(\mathcal{W}) = \mathcal{F}_n \mathcal{G}_n(\mathcal{W}) +$$
$$\sum_{m=1}^{n} \frac{(n-m)!}{n!\, m!} \mathcal{F}_n(\mathcal{W}) \underbrace{\vdots (\bar{\partial} \, K \, \vec{\partial}) \cdots (\bar{\partial} \, K \, \vec{\partial})}_{m\ factors} \vdots \mathcal{G}_n(\mathcal{W}). \qquad (9.150)$$

Here we have introduced the ordering $\vdots \cdots \vdots$, in which $\bar{\partial}_{\mathcal{W}_{a_i}}$ $\left(\bar{\partial}_{\mathcal{W}_{b_i}}\right)$ are moved to the left (right) extreme and $\bar{\partial}_{\mathcal{W}_{a_i}}$'s $\left(\bar{\partial}_{\mathcal{W}_{b_i}}\right)$'s act on everything to their left (right). In doing so one always has to remember to include the overall factor coming from graded commutations. Thus for example, $\vdots(\bar{\partial} \, K \, \vec{\partial})(\bar{\partial} \, K \, \vec{\partial})\vdots = (-1)^{|a||c|+|b|(|c|+|d|)} \bar{\partial}_{\mathcal{W}_a} \bar{\partial}_{\mathcal{W}_c} K_{ab} K_{cd} \bar{\partial}_{\mathcal{W}_b} \bar{\partial}_{\mathcal{W}_d}$. From (9.150) it is apparent that, in the graded commutative limit ($n \to \infty$), we get back the ordinary point-wise multiplication $\mathcal{F}_n \mathcal{G}_n(\mathcal{W})$. This formula was first derived in [9].

A consequence of (9.146) is the graded commutator of the $*$-product

$$[\mathcal{W}_a, \mathcal{W}_b\}_{*_n} = \frac{i}{n} f_{abc} \mathcal{W}_c \qquad (9.151)$$

which generalizes a familiar result for the usual $*$-products.

A special case of our result for the $*$-product follows if we restrict ourselves to the even subspace S_F^2 of $S_F^{(2,2)}$, namely the fuzzy sphere. In this case, $\mathcal{F}_n(\mathcal{W})$ and $\mathcal{G}_n(\mathcal{W})$ become $\mathcal{F}_n(\vec{x})$ and $\mathcal{G}_n(\vec{x})$ and we get from (9.150):

$$\mathcal{F}_n *_n \mathcal{G}_n(\vec{x}) = \mathcal{F}_n \mathcal{G}_n(\vec{x}) + \sum_{m=1}^{n} \frac{(n-m)!}{n!\, m!} \partial_{i_1} \cdots \partial_{i_m} \mathcal{F}_n(\vec{x})$$

$$\times K_{i_1 j_1} \cdots K_{i_m j_m} \partial_{j_1} \cdots \partial_{j_m} \mathcal{G}_n(\vec{x}), \qquad (9.152)$$

which is the formula given in 3.101.

9.7.2 The $*$-Product on Fuzzy "Sections of Bundles"

Let us first remark that the left- and right-actions of $\Psi_\mu^{L,R}$ and $(\Psi_\mu^\dagger)^{L,R}$ on $Mat(n+1,n)$ are defined and change n by an increment of 1:

$$\Psi_\mu^{L,R} Mat(n+1,n): \qquad \tilde{\mathcal{H}}_n \to \tilde{\mathcal{H}}_{n-1},$$

$$(\Psi_\mu^{L,R})^\dagger Mat(n+1,n): \qquad \tilde{\mathcal{H}}_n \to \tilde{\mathcal{H}}_{n+1}. \qquad (9.153)$$

On $|\psi', n\rangle$, we find

$$S_\mu |\psi', n\rangle = \psi'_\mu |\psi', n-1\rangle, \quad \langle \psi', n| S_\mu^\dagger = \langle \psi', n-1| \bar{\psi}_\mu'. \qquad (9.154)$$

Thus we get the matrix elements

$$\langle \psi', n-1| S_\mu |\psi', n\rangle = \psi'_\mu, \quad \langle \psi', n| S_\mu^\dagger |\psi', n-1\rangle = \bar{\psi}_\mu'. \qquad (9.155)$$

We observe that the r.h.s. of the equations in (9.155) define functions on $S^{(3,2)}$. Thus these matrix elements correspond to fuzzy sections of bundles on $S_F^{(2,2)}$. It is possible to obtain the $*$-product for these fuzzy sections of bundles. The results below also provide an alternative way to compute the $*$-products in (9.146) and (9.150).

For the $*$-product of ψ' with $\bar{\psi}'$ we find from

$$\langle \psi', n| S_\mu S_\nu^\dagger |\psi^{\dagger\prime}, n\rangle = \langle \psi', n| (-1)^{|S_\mu||S_\nu|} \frac{n}{n+1} S_\nu^\dagger S_\mu + \frac{1}{n+1} \delta_{\mu\nu} |\psi', n\rangle, \qquad (9.156)$$

that

$$\psi'_\mu * \bar{\psi}'_\nu = \frac{n}{n+1} \psi'_\mu \bar{\psi}'_\nu + \frac{1}{n+1} \delta_{\mu\nu}. \qquad (9.157)$$

[There is an abuse of notation here. $*$-products are for coordinate functions while (9.157) involves their values.] Here we have used (9.42) and the fact that $\psi'_\mu \bar{\psi}'_\nu = (-1)^{|S_\mu||S_\nu|} \bar{\psi}'_\nu \psi'_\mu$ to get rid of $(-1)^{|S_\mu||S_\nu|}$. Rearranging the last result we can write

$$\psi'_\mu * \bar{\psi}'_\nu = \frac{1}{n+1} \Omega_{\mu\nu} + \psi'_\mu \bar{\psi}'_\nu \,,$$

$$\Omega_{\mu\nu} \equiv \delta_{\mu\nu} - \psi'_\mu \bar{\psi}'_\nu \,. \tag{9.158}$$

The significance of $\Omega_{\mu\nu}$ will be be discussed shortly. Before that, as a check of our results of the previous section, we can compute $\mathcal{W}_a *_n \mathcal{W}_b$, using the method above. First note that

$$\mathcal{W}_a(\psi', \bar{\psi}', n) = \bar{\psi}' \Lambda_a^{(\frac{1}{2})} \psi' = \langle \psi', n | S^\dagger \Lambda_a^{(\frac{1}{2})} S | \psi', n \rangle \,. \tag{9.159}$$

Hence

$$\mathcal{W}_a *_n \mathcal{W}_b(\psi', \bar{\psi}', n)$$
$$= \langle \psi', n | S^\dagger_\mu (\Lambda_a^{(\frac{1}{2})})_{\mu\nu} S_\nu S^\dagger_\alpha (\Lambda_b^{(\frac{1}{2})})_{\alpha\beta} S_\beta | \psi', n \rangle$$
$$= \bar{\psi}'_\mu (\Lambda_a^{(\frac{1}{2})})_{\mu\nu} \left(\frac{1}{n} \Omega_{\nu\alpha} + \psi'_\nu \bar{\psi}'_\alpha \right) (\Lambda_b^{(\frac{1}{2})})_{\alpha\beta} \psi'_\beta$$
$$= \bar{\psi}'_\mu (\Lambda_a^{(\frac{1}{2})})_{\mu\nu} \left(\frac{1}{n} \delta_{\nu\alpha} + \frac{n-1}{n} \psi'_\nu \bar{\psi}'_\alpha \right) (\Lambda_b^{(\frac{1}{2})})_{\alpha\beta} \psi'_\beta \tag{9.160}$$
$$= \frac{1}{n} \mathcal{W}_a *_1 \mathcal{W}_b(\psi', \bar{\psi}', n) + \frac{n-1}{n} \mathcal{W}_a(\psi', \bar{\psi}', n) \mathcal{W}_b(\psi', \bar{\psi}', n)$$

which is (9.146).

Comparing the second line of the last equation with (9.148) we get the important result

$$K_{ab} = (\mathcal{W}_a \overleftarrow{\partial}_\mu) \Omega_{\mu\nu} (\overrightarrow{\partial}_\nu \mathcal{W}_b)$$
$$\equiv \mathcal{W}_a \overleftarrow{\partial} \, \Omega \, \overrightarrow{\partial} \mathcal{W}_b \,, \tag{9.161}$$

where $\overleftarrow{\partial} \, \Omega \, \overrightarrow{\partial} \equiv \overleftarrow{\partial}_\mu \Omega_{\mu\nu} \overrightarrow{\partial}$ and $\partial_\mu = \frac{\partial}{\partial \psi'_\mu}$.

We would like to note that this result can be used to write (9.150) in terms of $\overleftarrow{\partial} \, \Omega \, \overrightarrow{\partial}$. To this end we write

$$\mathcal{F}_n *_n \mathcal{G}_n(\mathcal{W}) = (-1)^{\sum_{j>i} |a_j||b_i|} F^{a_1 a_2 \cdots a_n} \prod_i (\mathcal{W}_{a_i}(1 + \overleftarrow{\partial} \, \Omega \, \overrightarrow{\partial}) \mathcal{W}_{b_i}) G^{b_1 b_2 \cdots b_n} \,. \tag{9.162}$$

Carrying out a similar calculation that lead to (9.150), one finally finds

$$\mathcal{F}_n *_n \mathcal{G}_n(\mathcal{W}) = \mathcal{F}_n \mathcal{G}_n(\mathcal{W}) +$$
$$\sum_{m=1}^n \frac{(n-m)!}{n!m!} \mathcal{F}_n(\mathcal{W}) \underbrace{: (\overleftarrow{\partial} \, \Omega \, \overrightarrow{\partial}) \cdots (\overleftarrow{\partial} \, \Omega \, \overrightarrow{\partial}) :}_{m\, factors} \mathcal{G}_n(\mathcal{W}) \,, \tag{9.163}$$

where now $\vdots \cdots \vdots$ takes $\overleftarrow{\partial}$ and $\overrightarrow{\partial}$ to the left and right extreme respectively. (When $\overleftarrow{\partial}$'s and $\overrightarrow{\partial}$'s are moved in this fashion, the phases coming from the graded commutators should be included just as for (9.150)).

It can be explicitly shown that $\Omega = (\Omega_{\mu\nu})$ is a projector, i.e.,

$$\Omega^2 = \Omega \quad \text{and} \quad \Omega^\ddagger = \Omega. \tag{9.164}$$

Due to (9.161), the last equation implies similar properties for [§]

$$\mathcal{K}_{ab} \equiv (K\,S^{-1})_{ab}. \tag{9.165}$$

which we discuss next.

9.8 More on the Properties of $\mathcal{K}_{ab}$

A closer look at the properties of $\mathcal{K}_{ab} \equiv (K\,S^{-1})_{ab}$, where

$$\begin{aligned}
K_{ab}(\psi) &= \mathcal{W}_a *_1 \mathcal{W}_b(\psi) - \mathcal{W}_a(\psi)\,\mathcal{W}_b(\psi) \\
&= \langle \psi', 1|\widehat{\mathcal{W}_a\mathcal{W}_b}|\psi', 1\rangle - \langle \psi', 1|\widehat{\mathcal{W}_a}|\psi', 1\rangle\langle \psi', 1|\widehat{\mathcal{W}_b}|\psi', 1\rangle, \quad (9.166)
\end{aligned}$$

will give us more insight on the structure of the $*$-product found in the previous section. First note that $\mathcal{K}_{ab}$ depends on both ψ and $\bar{\psi}$. We denote this dependence by $\mathcal{K}_{ab}(\psi)$ for short, omitting to write the $\bar{\psi}$ dependence. Now we would like to show that the matrix $\mathcal{K}(\psi) = (\mathcal{K}_{ab}(\psi))$ is a projector.

We first recall that the $\left(\frac{1}{2}\right)_{osp(2,2)+}$, representation of $osp(2,2)$ is at the same time the $J_{osp(2,1)} = \left(\frac{1}{2}\right)_{osp(2,1)}$ irreducible representation of $osp(2,1)$. Their highest and lowest weight states are given by

$$|J_{osp(2,1)}, J_{su(2)}, J_3\rangle = \begin{cases} |\frac{1}{2}, \frac{1}{2}, \frac{1}{2}\rangle & \equiv |\text{highest weight state}, \\ |\frac{1}{2}, \frac{1}{2}, -\frac{1}{2}\rangle & \equiv |\text{lowest weight state} \end{cases} \tag{9.167}$$

We note that, starting from the lowest weight state $|1/2, 1/2, -1/2\rangle = \Psi_2^\dagger|0\rangle$, one can construct another supercoherent state, expressed by a formula similar to (9.84). Now consider $\mathcal{W}_a^\pm(\psi^0)$ at $\psi = \psi^0 = (1, 0, 0)$ obtained from computing $\widehat{\mathcal{W}}_a$ in the supercoherent states induced from the states given in (9.167):

$$\mathcal{W}^\pm(\psi^0) = (\mathcal{W}_1(\psi^0)\cdots\mathcal{W}_8(\psi^0)) = \left(0, 0, \pm\frac{1}{2}, 0, 0, 0, 0, 1\right). \tag{9.168}$$

[§]We consider all the indices down through out this chapter. In the following section the relevant object under investigation is $\mathcal{K}_{ab}$ corresponding to $K_a{}^b$ in a notation where indices are raised and lowered by the metric.

In (9.168) $+(-)$ corresponds to upper(lower) entries in (9.167) and the calculation is done using (9.50) and (9.60).

Although not essential in what follows, we remark that $\mathcal{W}^-(\psi = (1,0,0)) = \mathcal{W}^+(\psi = (0,1,0))$, that is,

$$\mathcal{W}_a^-(\psi^0) = \mathcal{W}_b^+(\psi^0)\,(Ad\,e^{i\pi\Lambda_2^{(\frac{1}{2})}})_{ba}\,. \tag{9.169}$$

Note that all other points in $S_F^{(2,2)}$ can be obtained from $\mathcal{W}^\pm(\psi^0)$ by the adjoint action of the group, i.e.,

$$\mathcal{W}_a^\pm(\psi) = \mathcal{W}_b^\pm(\psi^0)(Ad\,g^{-1})_{ba} \tag{9.170}$$

where $\psi = g\psi^0$.

We define $\mathcal{K}^\pm(\psi^0)$ using $\mathcal{W}^\pm(\psi^0)$ for $\mathcal{W}$, and the equations (9.165), (9.166). The matrices $\mathcal{K}^\pm(\psi^0)$ when computed at the fiducial points (using for instance (9.24), (9.145), (9.147)) have the block diagonal forms

$$\mathcal{K}^\pm(\psi^0) = (\mathcal{K}_{ab}^\pm(\psi^0)) =$$
$$\begin{pmatrix} \left(\frac{1}{2}\delta_{ij} \pm \frac{i}{2}\epsilon_{ij3} - 2\,\mathcal{W}_i^\pm(\psi^0)\,(\mathcal{W}_j^\pm(\psi^0))\right)_{3\times3} & 0 & 0 \\ 0 & \left(\Sigma_{\alpha\beta}^\pm\right)_{4\times4} & 0 \\ 0 & 0 & 0 \end{pmatrix} \tag{9.171}$$

with

$$\Sigma^\pm = (\Sigma_{\alpha\beta}^\pm) = \frac{1}{4}\begin{pmatrix} 1\pm\sigma_3 & -(1\pm\sigma_3) \\ -(1\pm\sigma_3) & 1\pm\sigma_3 \end{pmatrix} \tag{9.172}$$

where the upper (lower) sign stands for the upper (lower) sign in $\mathcal{W}^\pm(\psi^0)$. The supermatrices $\mathcal{K}^\pm(\psi^0)$ are even and consequently do not mix the $1,2,3,8$ and $4,5,6,7$ entries of a (super)vector. Its grade adjoint is its ordinary adjoint †. Now from (9.171), it is straightforward to check that the relations

$$\begin{aligned} (\mathcal{K}^\pm(\psi^0))^2 &= \mathcal{K}^\pm(\psi^0)\,, \\ (\mathcal{K}^\pm(\psi^0))^\ddagger &= \mathcal{K}^\pm(\psi^0)\,, \\ \mathcal{K}^+(\psi^0)\,\mathcal{K}^-(\psi^0) &= 0 \end{aligned} \tag{9.173}$$

are fulfilled. (9.173) establishes that $\mathcal{K}^\pm(\psi^0)$ are orthogonal projectors. By the adjoint action of the group, we have

$$\mathcal{K}_{ab}^\pm(\psi) = ((Ad\,g)^T)_{ad}^{-1}\,\mathcal{K}_{de}^\pm(\psi^0)\,(Ad\,g)_{eb}^T\,, \tag{9.174}$$

with T denoting the transpose. (9.174) implies that $\mathcal{K}^\pm(\psi)$ are projectors for all $g \in OSp(2,2)$.

We further observe that a super-analogue $\mathcal{J}$ of the complex structure can be defined over the supersphere. To show this, we first observe that the projective module for "sections of the supertangent bundle" $TS^{(2,2)}$ over $S^{(2,2)}$ is $\mathcal{P}\mathcal{A}^8$, where $\mathcal{A}$ is the algebra of superfunctions over $S^{(2,2)}$, $\mathcal{A}^8 = \mathcal{A} \otimes_{\mathbb{C}} \mathbb{C}^8$ and

$$\mathcal{P}(\psi) = \mathcal{K}^+(\psi) + \mathcal{K}^-(\psi) \tag{9.175}$$

is a projector. The super-complex structure is the operator with eigenvalues $\pm i$ on the subspaces $TS^{(2,2)}_{\pm}$ of $TS^{(2,2)}$ with $\mathcal{K}^{\pm}\mathcal{A}^8 \equiv TS^{(2,2)} = TS^{(2,2)}_{+} \oplus TS^{(2,2)}_{-}$. It is given by the matrix $\mathcal{J}$ with elements

$$\mathcal{J}_{ab}(\psi) = -i(\mathcal{K}^+ - \mathcal{K}^-)_{ab}(\psi), \tag{9.176}$$

and acts on $\mathcal{P}\mathcal{A}^8$. Since

$$\mathcal{J}^2(\psi)\Big|_{\mathcal{P}\mathcal{A}^8} = -\mathcal{P}(\psi)\Big|_{\mathcal{P}\mathcal{A}^8} = -\mathbf{1}\Big|_{\mathcal{P}\mathcal{A}^8} \tag{9.177}$$

$\left(\delta\Big|_{\varepsilon}\right.$ denoting the restriction of δ to ε), it indeed defines a super complex structure. Furthermore, due to the relation

$$\mathcal{J}\Big|_{\mathcal{K}^{\pm}\mathcal{A}^8} = \mp i\Big|_{\mathcal{K}^{\pm}\mathcal{A}^8}, \tag{9.178}$$

$\mathcal{K}^{\pm}\mathcal{A}^8$ give the "holomorphic" and "anti-holomorphic" parts of $\mathcal{P}\mathcal{A}^8$. Finally, we can also write

$$\mathcal{K}^{\pm}(\psi) = \frac{1}{2}(-\mathcal{J}^2 \pm i\mathcal{J})(\psi). \tag{9.179}$$

9.9 The $O(3)$ Nonlinear Sigma Model on $S^{(2,2)}$

As a final topic in this chapter, we describe the "$O(3)$ nonlinear SUSY sigma model" on $S^{(2,2)}$ and $S_F^{(2,2)}$. We follow the discussion in [123].

9.9.1 *The Model on* $S^{(2,2)}$

On $S^{(2,2)}$, it is defined by the action

$$\mathcal{S}^{SUSY} = -\frac{1}{4\pi}\int d\mu \left(C_{\alpha\beta}d_{\alpha}\Phi^a d_{\beta}\Phi^a + \frac{1}{4}\gamma\Phi^a\gamma\Phi^a\right), \tag{9.180}$$

where $\Phi^a = \Phi^a(x_i, \theta_\alpha)$, $(a = 1, 2, 3)$ is a real triplet superfield fulfilling the constraint

$$\Phi^a\Phi^a = 1, \quad (a = 1, 2, 3). \tag{9.181}$$

Obviously, the world sheet for this theory is $S^{(2,2)}$ while the target manifold is a 2-sphere.

A closely related model is the one formulated on the standard $(2,1)$-dimensional superspace $\mathcal{C}^{(2,1)}$, first studied by Witten, and by Di Vecchia et al.[124, 125].

The triplet superfield Φ^a can be expanded in powers of θ_α as

$$\Phi^a(x_i\,,\theta_\alpha) = n^a(x_i) + C_{\alpha\beta}\theta_\beta\psi_\alpha^a(x_i) + \frac{1}{2}F^a(x_i)C_{\alpha\beta}\theta_\alpha\theta_\beta \qquad (9.182)$$

where $\psi^a(x_i)$ are two component Majorana spinors : $\psi_\alpha^{a\dagger} = C_{\alpha\beta}\psi_\beta^a$, and $F^a(x_i)$ are auxiliary scalar fields. In terms of the component fields, the constraint equation (9.181) splits into

$$n^a n^a = 1\,, \qquad (9.183a)$$

$$n^a F^a = \frac{1}{2}\psi^{a\dagger}\psi^a\,, \qquad (9.183b)$$

$$n^a \psi_\alpha^a = 0\,. \qquad (9.183c)$$

(9.183a) is the usual constraint of $O(3)$ non-linear sigma model defined earlier in chapter 6 by the action [66]

$$S = -\frac{1}{8\pi}\int_{S^2} d\Omega(\mathcal{L}_i n_a)(\mathcal{L}_i n_a)\,. \qquad (9.184)$$

Thus, we see that bosonic sector of the S^{SUSY} coincides with the $\mathbb{C}P^1$ sigma model. The other two constraints are additional. We note that (9.183b) can be used along with the equations of motion for F^a to eliminate F^a's from the action. The techniques for performing such calculations can be found for instance in [125].

9.9.2   The Model on $S_F^{(2,2)}$

The fuzzy action approaching (9.180) for large n is [123]

$$\mathcal{S}^{SUSY} = str\Big(C_{\alpha\beta}\,[D_\alpha\,,\hat{\Phi}^a\}\,[D_\beta\,,\hat{\Phi}^a\} + \frac{1}{4}[\Gamma\,,\hat{\Phi}^a]\,[\Gamma\,,\hat{\Phi}^a]\Big)\,, \qquad (9.185)$$

where

$$\hat{\Phi}^a\hat{\Phi}^a = \mathbf{1}_{2n+1}\,, \quad \mathbf{1}_{2n+1} \in Mat(n+1,n)\,. \qquad (9.186)$$

(9.186) can be expressed in terms of the $*$-product on $S_F^{(2,2)}$ as

$$\Phi^a * \Phi^a(\psi',\bar{\psi}',n) = 1\,. \qquad (9.187)$$

This expression involves the product of derivatives of Φ^a up to n^{th} order, and is not easy to work with. Alternatively we can construct supersymmetric extensions of "Bott Projectors" introduced in chapter 6 to study this model, as we indicate below.

9.9.3 *Supersymmetric Extensions of Bott Projectors*

A possible supersymmetric extension of the projector $\mathcal{P}_\kappa(x)$ can be obtained in the following manner. Let $\mathcal{U}(x_i, \theta_\alpha)$ be a graded unitary operator :

$$\mathcal{U}\mathcal{U}^\ddagger = \mathcal{U}^\ddagger\mathcal{U} = 1. \tag{9.188}$$

$\mathcal{U}(x_i, \theta_\alpha)$ can be thought as a 2×2 supermatrix whose entries are functions on $S^{(2,2)}$. $\mathcal{U}(x_i, \theta_\alpha)$ acts on $\mathcal{P}_\kappa$ by conjugation and generates a set of supersymmetric projectors $\mathcal{Q}_\kappa(x_i, \theta_\alpha)$:

$$\mathcal{Q}_\kappa(x_i, \theta_\alpha) = \mathcal{U}^\ddagger \mathcal{P}_\kappa(x)\mathcal{U}. \tag{9.189}$$

It is easy to see that $\mathcal{Q}_\kappa(x_i, \theta_\alpha)$ satisfies

$$\mathcal{Q}_\kappa^2(x_i, \theta_\alpha) = Q_\kappa(x_i, \theta_\alpha), \quad \text{and} \quad \mathcal{Q}_\kappa^\ddagger(x_i, \theta_\alpha) = \mathcal{Q}_\kappa(x_i, \theta_\alpha). \tag{9.190}$$

Thus $\mathcal{Q}_\kappa(x_i, \theta_\alpha)$ is a (super-)projector. The real superfields on $S^{(2,2)}$ associated to $\mathcal{Q}_\kappa(x_i, \theta_\alpha)$ are given by

$$\Phi'_a(x_i, \theta_\alpha) = Tr\,\tau_a \mathcal{Q}_\kappa. \tag{9.191}$$

In order to check that $\mathcal{Q}_\kappa(x_i, \theta_\alpha)$ reproduces the superfields on $S^{(2,2)}$ subject to

$$\Phi'_a\Phi'_a = 1, \tag{9.192}$$

we proceed as follows. First we expand $\mathcal{U}(x_i, \theta_\alpha)$ in powers of Grassmann variables as

$$\mathcal{U}(x_i, \theta_\alpha) = \mathcal{U}_0(x_i) + C_{\alpha\beta}\theta_\beta\mathcal{U}_\alpha(x_i) + \frac{1}{2}\mathcal{U}_2(x_i)C_{\alpha\beta}\theta_\alpha\theta_\beta \tag{9.193}$$

where $\mathcal{U}_0, \mathcal{U}_\alpha(\alpha = \pm)$ and $\mathcal{U}_2$ are all 2×2 graded unitary matrices. The requirement of graded unitarity for $\mathcal{U}(x_i, \theta_\alpha)$ implies the following for the component matrices:

i. $\mathcal{U}_0(x_i)$ is unitary,

ii. $\mathcal{U}_\alpha(x_i)$ are uniquely determined by

$$\mathcal{U}_\alpha(x_i) = H_\alpha(x_i)\mathcal{U}_0(x_i), \tag{9.194}$$

where H_α are 2×2 odd supermatrices satisfying the reality condition $H_\alpha^\ddagger = -C_{\alpha\beta}H_\beta$,

iii. $\mathcal{U}_2$ is of the form $\mathcal{U}_2 = A\mathcal{U}_0$ with A being an 2×2 even supermatrix, whose symmetric part satisfies

$$A + A^\dagger = -C_{\alpha\beta}H_\alpha H_\beta. \tag{9.195}$$

Using (9.193) in (9.189) and the conditions listed above, we can extract the component fields of the superfield $\Phi'_a(x_i, \theta_\alpha)$. We find

$$n_a^{\kappa'} := Tr\, \tau_a U_0^\dagger \mathcal{P}_\kappa U_0\,, \tag{9.196}$$

$$\psi_\alpha^{a'} := Tr\, \tau_a U_0^\dagger [H_\alpha, \mathcal{P}_\kappa] U_0 = -2i(\vec{n}^{\kappa'} \times \vec{H}'_\alpha)^a\,, \tag{9.197}$$

and, after using (9.195),

$$F'_a := Tr\, \tau_a U_0^\dagger (\mathcal{P}_\kappa A + A^\dagger \mathcal{P}_\kappa - C_{\alpha\beta} H_\beta \mathcal{P}_\kappa H_\alpha) U_0 \tag{9.198}$$

$$= 4(\vec{H}'_+ \cdot \vec{H}'_-)n_a^{\kappa'} - 2\vec{H}_+^{a'}(\vec{n}^{\kappa'} \cdot \vec{H}'_-)$$

$$-(\vec{n}^{\kappa'} \cdot \vec{H}'_+)2\vec{H}_-^{a'} + i(\vec{n}^{\kappa'} \times (\vec{A}' - \vec{A}^{\dagger'}))^a\,,$$

where $\vec{H}'_\alpha = H_\alpha^{1'}\tau^1 + H_\alpha^{2'}\tau^2$ and $\vec{A}' = A^{3'}\tau^3$. By direct computation from above it follows that

$$n_a^{\kappa'}n_a^{\kappa'} = 1\,, \qquad n_a^{\kappa'}F'_a = \frac{1}{2}\psi_a^{\dagger'}\psi'_a\,, \qquad n_a^{\kappa'}\psi_\pm^{a'} = 0\,. \tag{9.199}$$

Comparing (9.199) with (9.183) we observe that they are identical. Therefore, we conclude that the superfield associated to the super-projector $\mathcal{Q}_\kappa$ is the same as the superfield of the supersymmetric nonlinear sigma model discussed previously.

9.9.4 *The SUSY Action Revisited*

We now extend (9.129) by including winding number sectors.

Equipped with the supersymmetric projector $\mathcal{Q}_\kappa$ we can write, in close analogy with the $\mathbb{C}P^1$ model, the action for the supersymmetric nonlinear $O(3)$ sigma model for winding number κ as

$$S_\kappa^{SUSY} = -\frac{1}{2\pi} \int d\mu\, Tr\Big[C_{\alpha\beta}\,(d_\alpha \mathcal{Q}_\kappa)(d_\beta \mathcal{Q}_\kappa) + \frac{1}{4}(\gamma \mathcal{Q}_\kappa)(\gamma \mathcal{Q}_\kappa)\Big]. \tag{9.200}$$

The even part of this action, as well as the one given in (9.180) are nothing but the action S_κ of the $\mathbb{C}P^1$ theory given in (6.22) and (9.184), respectively. In other words, the action S_κ^{SUSY} is the supersymmetric extension of S_κ on S^2 to $S^{(2,2)}$. Consequently, in the supersymmetric theory, it is possible to interpret the index κ carried by the action as the winding number of the corresponding $\mathbb{C}P^1$ theory. For $\kappa = 0$ we get back (9.129).

We recall that d_α and γ are both graded derivations in the superalgebra $osp(2,2)$. Therefore, they obey a graded Leibniz rule. From $\mathcal{Q}_\kappa^2 = \mathcal{Q}_\kappa$, we find

$$\mathcal{Q}_\kappa d_\alpha \mathcal{Q}_\kappa = d_\alpha \mathcal{Q}_\kappa(1 - \mathcal{Q}_\kappa)\,. \tag{9.201}$$

This enables us to write

$$Tr d_\alpha \mathcal{Q}_\kappa (\mathbf{1} - \mathcal{Q}_\kappa) d_\alpha \mathcal{Q}_\kappa = Tr(\mathbf{1} - \mathcal{Q}_\kappa)(d_\alpha \mathcal{Q}_\kappa)^2 = \frac{1}{2} Tr(d_\alpha \mathcal{Q}_\kappa)^2 . \quad (9.202)$$

Equations (9.201) and (9.202) continue to hold when d_α is replaced by γ as well. The action can also be written as

$$\mathcal{S}_\kappa^{SUSY} = -\frac{1}{\pi} \int d\mu \, Tr \left[C_{\alpha\beta} \mathcal{Q}_\kappa (d_\alpha \mathcal{Q}_\kappa)(d_\beta \mathcal{Q}_\kappa) + \frac{1}{4} \mathcal{Q}_\kappa (\gamma \mathcal{Q}_\kappa)(\gamma \mathcal{Q}_\kappa) \right] . \quad (9.203)$$

9.9.5 *Fuzzy Projectors and Sigma Models*

In much the same way that the supersymmetric projectors $\mathcal{Q}_\kappa$ have been constructed from $\mathcal{P}_\kappa$ in the previous section, we can construct the supersymmetric extensions of $\widehat{\mathcal{P}}_\kappa$ by the graded unitary transformation

$$\widehat{\mathcal{Q}}_\kappa = \widehat{\mathcal{U}}^\ddagger \widehat{\mathcal{P}}_\kappa \widehat{\mathcal{U}} \quad (9.204)$$

where now $\widehat{\mathcal{U}}$ is a 2×2 supermatrix whose entries are polynomials in not only $a_\alpha^\dagger a_\beta$, but also in $b^\dagger b$. The domain of $\mathcal{U}_{ij}$ is $\tilde{\mathcal{H}}_n$.

$\widehat{\mathcal{Q}}_\kappa$ acts on the finite-dimensional space $\tilde{\mathcal{H}}_n^2 = \tilde{\mathcal{H}}_n \otimes \mathbb{C}^2$. We can check that

$$[\widehat{\mathcal{Q}}_\kappa , \widehat{N}\} = 0 , \quad (9.205)$$

where $\widehat{N} = a_\alpha^\dagger a_\alpha + b^\dagger b$ is the number operator on $\tilde{\mathcal{H}}_n$. In close analogy with the fuzzy $\mathbb{C}P^1$ model, it is now possible to write down a finite-dimensional (super-)matrix model for the (super-)projectors $\widehat{\mathcal{Q}}_\kappa$.

The action for the fuzzy supersymmetric model becomes

$$S_{F,\kappa}^{SUSY} = \frac{1}{2\pi} \, Str_{\widehat{N}=n} \left(C_{\alpha\beta} [D_\alpha , \widehat{\mathcal{Q}}_\kappa\} [D_\beta , \widehat{\mathcal{Q}}_\kappa\} + \frac{1}{4} [\Gamma , \widehat{\mathcal{Q}}_\kappa] [\Gamma , \widehat{\mathcal{Q}}_\kappa] \right) . \quad (9.206)$$

Str in the above expression is the supertrace over $\tilde{\mathcal{H}}_n^2$. In the large $\widehat{N} = n$ limit, (9.206) approximates the action given in (9.200).

This concludes our discussion of the nonlinear sigma model on $S_F^{(2,2)}$.

Chapter 10

SUSY Anomalies on the Fuzzy Supersphere

10.1 Overview

In this section, we give an overview of fuzzy SUSY for convenience. Much of it is in chapter 9 also. The notation for irreducible representations is different from that of chapter 9.

10.1.1 *The Fuzzy Sphere*

We recall that the fuzzy sphere $S_F^2(n)$ is the $(n + 1) \times (n + 1)$ matrix algebra $Mat(n + 1)$. It can be realized as linear operators on $\mathcal{H}^{n+1}$ with the orthonormal basis vectors

$$\frac{(a_1^\dagger)^{n_1}}{\sqrt{n_1!}} \frac{(a_2^\dagger)^{n_2}}{\sqrt{n_2!}} |0\rangle \ , \quad n_1 + n_2 = n \ , \tag{10.1}$$

where a_i, $a_i^\dagger$ are bosonic oscillators. The vectors (10.1) span a subspace of the Fock space with fixed particle number n:

$$N := \sum_i a_i^\dagger a_i \ , \quad N|_{\mathcal{H}^{n+1}} = n \ . \tag{10.2}$$

In this representation, the elements of $S_F^2(n)$ are the linear operators

$$\sum_{i,j} c_{i,j}^m (a_i^\dagger)^m (a_j)^m \ , \quad c_{i,j}^m \in \mathbb{C} \ , \tag{10.3}$$

restricted to the subspace $\mathcal{H}^{n+1}$.

The group $SU(2)$ acts on $\mathcal{H}^{n+1}$ and hence on $S_F^2(n)$ by its spin $\frac{n}{2}$ unitary irreducible representation. The angular momentum generators are

$$L_i = a^\dagger \frac{\sigma_i}{2} a \ , \quad \sigma_i \text{ are Pauli matrices.} \tag{10.4}$$

10.1.2 *SUSY*

The $\mathcal{N} = 1$ SUSY version of $SU(2)$ is $OSp(2,1)$. It has the graded Lie algebra $osp(2,1)$. Its generators (basis) can be written using oscillators if we introduce one additional fermionic oscillator b and its adjoint $b^\dagger$. They commute with a_i, $a_j^\dagger$. Then the $osp(2,1)$ generators are

$$\Lambda_i = a^\dagger \frac{\sigma_i}{2} a \ , \quad \Lambda_4 = -\frac{1}{2}(a_1^\dagger b + b^\dagger a_2) \ , \tag{10.5}$$

$$\Lambda_5 = \frac{1}{2}(-a_2^\dagger b + b^\dagger a_1) \ , \quad \sigma_i = \text{Pauli matrices.}$$

The $\mathcal{N} = 2$ SUSY version of $SU(2)$ is $OSp(2,2)$. It has the graded Lie algebra $osp(2,2)$. Its basis consists of the $osp(2,1)$ generators and three additional generators

$$\Lambda_{4'} \equiv \Lambda_6 = \frac{1}{2}(a_1^\dagger b - b^\dagger a_2) \ , \quad \Lambda_{5'} \equiv \Lambda_7 = \frac{1}{2}(a_2^\dagger b + b^\dagger a_1) \ , \tag{10.6}$$

$$\Lambda_8 = a^\dagger a + 2b^\dagger b \ .$$

If $\{\cdot, \cdot\}$ denotes the anticommutator, $osp(2,2)$ has the defining relations

$$[\Lambda_i, \Lambda_j] = i\varepsilon_{ijk}\Lambda_k \ , \quad [\Lambda_i, \Lambda_\alpha] = \frac{1}{2}\Lambda_\beta(\sigma_i)_{\beta\alpha} \ , \quad \{\Lambda_\alpha, \Lambda_\beta\} = \frac{1}{2}(\varepsilon\sigma_i)_{\alpha\beta}\Lambda_i \ ,$$

$$[\Lambda_i, \Lambda_8] = 0 \ , \quad [\Lambda_\alpha, \Lambda_8] = -\Lambda_{\alpha'} \ , \quad \{\Lambda_\alpha, \Lambda_{\alpha'}\} = \frac{1}{4}\varepsilon_{\alpha\beta}\Lambda_8 \ ,$$

$$\{\Lambda_{\alpha'}, \Lambda_{\beta'}\} = -\frac{1}{2}(\varepsilon\sigma_i)_{\alpha\beta}\Lambda_i \ , \quad [\Lambda_{\alpha'}, \Lambda_8] = -\Lambda_\alpha \ , \quad \varepsilon = \begin{pmatrix} 0 & 1 \\ -1 & 0 \end{pmatrix} . \tag{10.7}$$

These relations show in particular that the additional three generators form a triplet under $osp(2,1)$.

Conventional Lie algebras like that of $su(2)$ have a $*$ or an adjoint operation † defined on them. For Λ_i, it is just $\Lambda_i^\dagger = \Lambda_i$. This follows from the fact that $a_i^\dagger$ is the adjoint of a_i. For $osp(2,1)$ and $osp(2,2)$, † is replaced by the grade adjoint ‡. On the oscillators, ‡ is defined by

$$a_i^\ddagger = a_i^\dagger \ , \quad (a_i^\dagger)^\ddagger = (a_i^\dagger)^\dagger = a_i \ , \quad b^\ddagger = b^\dagger \ , \quad (b^\ddagger)^\ddagger = -b \ . \tag{10.8}$$

Hence $\ddagger = \dagger$ on bosonic oscillators.

On products of operators, ‡ is defined as follows. We assign the grade 0 to a_i, $a_j^\dagger$ and their products and 1 to b and $b^\dagger$. The grades are additive (mod 2). The grade of an operator L with definite grade is denoted by $|L|$. Then if L, M have definite grades, $(LM)^\ddagger \equiv (-1)^{|L||M|}M^\ddagger L^\ddagger$. Hence $(b^\dagger b)^\ddagger = b^\dagger b$ and

$$\Lambda_i^\ddagger = \Lambda_i \ , \quad \Lambda_\alpha^\ddagger = -\varepsilon_{\alpha\beta}\Lambda_\beta \ , \quad \Lambda_{\alpha'}^\ddagger = \varepsilon_{\alpha\beta}\Lambda_{\beta'} \quad \Lambda_8^\ddagger = \Lambda_8 \ . \tag{10.9}$$

10.1.3 Irreducible Representations

Let $osp(2,0)$ denote $su(2)$, the Lie algebra of $SU(2)$. Its IRR's are Γ_J^0, $J \in \mathbb{N}/2$. (Here $\mathbb{N} = \{0, 1, 2, ...\}$.) J has the meaning of angular momentum.

The $osp(2,1)$ algebra is of rank 1 just as $osp(2,0)$. We can take Λ_3 to be the generator of its Cartan subalgebra. Since

$$[\Lambda_3, \Lambda_4] = \frac{1}{2}\Lambda_4 \ , \quad [\Lambda_3, \Lambda_+ = \Lambda_1 + i\Lambda_2] = \Lambda_+ \ , \qquad (10.10)$$

Λ_4, Λ_+ are its raising operators. They commute:

$$[\Lambda_4, \Lambda_+] = 0 \ . \qquad (10.11)$$

In an IRR, both vanish on the highest weight vector. The eigenvalue $J \in \mathbb{N}/2$ of Λ_3 on the highest weight vector can be used to label its IRR's. They are denoted by Γ_J^1 in this chapter.

When restricted to its subalgebra $osp(2,0)$, Γ_J^1 splits as follows:

$$\Gamma_J^1\big|_{osp(2,0)} = \Gamma_J^0 \oplus \Gamma_{J-\frac{1}{2}}^0 \ , \quad J \geq \frac{1}{2} \ . \qquad (10.12)$$

Γ_0^1 is the trivial IRR.

The dimension of Γ_J^1 is $4J + 1$.

The graded Lie algebra $osp(2,2)$ is of rank 2. A basis for its Cartan subalgebra is $\{\Lambda_3, \Lambda_8\}$. Since

$$[\Lambda_3, \Lambda_4 + \Lambda_{4'}] = \frac{1}{2}(\Lambda_4 + \Lambda_{4'}) \ , \quad [\Lambda_8, \Lambda_4 + \Lambda_{4'}] = \Lambda_4 + \Lambda_{4'} \ , \qquad (10.13)$$

$\Lambda_4 + \Lambda_{4'}$ serves as the raising operator for both Λ_3 and Λ_8. We also have that $\Lambda_1 + i\Lambda_2 = \Lambda_+$ is the raising operator for Λ_3 alone:

$$[\Lambda_3, \Lambda_+] = \Lambda_+ \ , \quad [\Lambda_8, \Lambda_+] = 0 \ . \qquad (10.14)$$

The raising operators $\Lambda_4 + \Lambda_{4'}$ and Λ_+ commute:

$$[\Lambda_4 + \Lambda_{4'}, \Lambda_+] = 0 \ . \qquad (10.15)$$

Both vanish on the highest weight vector in an IRR while the eigenvalues $J \in \mathbb{N}/2$ and $k \in \mathbb{Z}$ of Λ_3 and Λ_8 on the highest weight vector can be used as labels of the IRR. They are denoted in this paper by $\Gamma_J^2(k)$.

The $osp(2,2)$ IRR's fall into classes, the *typical* and *atypical* (or *short*) IRR's. In the typical IRR's, $2|k| \neq J$ or $k = J = 0$, while in the atypical IRR's, $2|k| = J \neq 0$. The typical IRR with $2|k| \neq J$ restricted to $osp(2,1)$ splits as follows:

$$\Gamma_J^2(k)\big|_{osp(2,1)} = \Gamma_J^1 \oplus \Gamma_{J-\frac{1}{2}}^1 \ , \quad J \geq \frac{1}{2} \ . \qquad (10.16)$$

$\Gamma_0^2(0)$ is the trivial representation.

The atypical IRR's $\Gamma_J^2(\pm 2J)$ ($J \geq 1/2$) remain irreducible on restriction to $osp(2,1)$:

$$\Gamma_J^2(\pm 2J)\big|_{osp(2,1)} = \Gamma_J^1 \ . \tag{10.17}$$

$\Gamma_J^2(\pm J/2)$ can also be abbreviated to $\Gamma_{J\pm}^2$:

$$\Gamma_J^2(\pm 2J) \equiv \Gamma_{J\pm}^2 \ , \quad J \geq 1/2 \ . \tag{10.18}$$

$osp(2,2)$ admits the automorphism

$$\tau : \ \Lambda_i \to \Lambda_i \ , \quad \Lambda_\alpha \to \Lambda_\alpha \ , \quad \Lambda_{\alpha'} \to -\Lambda_{\alpha'} \ , \quad \Lambda_8 \to -\Lambda_8 \tag{10.19}$$

which interchanges $\Gamma_J^2(\pm k)$:

$$\tau : \ \Gamma_J^2(k) \to \Gamma_J^2(-k) \ . \tag{10.20}$$

10.1.4   *Casimir Operators*

The $osp(2,0) := su(2)$ Casimir operator K_0 is well-known:

$$K_0 = \Lambda_i^2 \ . \tag{10.21}$$

The $osp(2,1)$ Casimir operator is

$$K_1 = \Lambda_i^2 + \varepsilon_{\alpha\beta}\Lambda_\alpha\Lambda_\beta \ . \tag{10.22}$$

We have that

$$K_1\big|_{\Gamma_J^1} = J(J + \frac{1}{2})\mathbb{1} \ . \tag{10.23}$$

The $osp(2,2)$ quadratic Casimir operator is

$$K_2 = K_1 - \varepsilon_{\alpha\beta}\Lambda_{\alpha'}\Lambda_{\beta'} - \frac{1}{4}\Lambda_8^2 := K_1 - V_0 \ . \tag{10.24}$$

It has the property

$$K_2\big|_{\Gamma_J^2(k)} = J^2 - \frac{k^2}{4} \ ,$$
$$K_2\big|_{\Gamma_{J\pm}^2} = 0 \ . \tag{10.25}$$

As already mentioned, the IRR's $\Gamma_{J\pm}^2$ can be distinguished by the sign of Λ_8 on the highest weight vector.

$osp(2,2)$ also has a cubic Casimir operator [126], but we will not have occasion to use it.

10.1.5 Tensor Products

The basic Clebsch-Gordan series we need to know is as follows:

$$\Gamma^1_J \otimes \Gamma^1_K = \Gamma^1_{J+K} \oplus \Gamma^1_{J+K-1/2} \oplus \cdots \oplus \Gamma^1_{|J-K|} \ . \tag{10.26}$$

10.1.6 The Supertrace and the Grade Adjoint

Because of the decomposition (10.12), the vector space $\mathbb{C}^{4J+1}$ on which Γ^1_J acts can be written as $\mathbb{C}^{2J+1} \oplus \mathbb{C}^{2J}$ where the first term has angular momentum J and the second term has angular momentum $J - 1/2$. By definition, the first term is the even subspace and the second term is the odd subspace. The supertrace str of a matrix

$$M = \begin{pmatrix} P_{(2J+1)\times(2J+1)} & Q_{(2J+1)\times 2J} \\ R_{2J\times(2J+1)} & S_{2J\times 2J} \end{pmatrix} \tag{10.27}$$

is accordingly

$$str M = tr P - tr S \ . \tag{10.28}$$

The grade adjoint $M^\ddagger$ can be calculated using the rules of graded vector spaces (cf. chapter 9). The result is

$$M^\ddagger = \begin{pmatrix} P^\dagger & -R^\dagger \\ Q^\dagger & S^\dagger \end{pmatrix} \tag{10.29}$$

This formula is coherent with (10.9).

If $Q, R = 0$, we say that M is even, while if $P, S = 0$, we say that M is odd. We assign a number $|M| = 0, 1 \pmod 2$ to even and odd matrices M respectively.

10.1.7 The Free Action

The space with $N = n$ has maximum angular momentum $J = n/2$. It carries the $osp(2,1)$ IRR $\Gamma^1_{n/2}$ which splits under $su(2)$ into $\Gamma^0_{n/2} \oplus \Gamma^0_{(n-1)/2}$. It carries either of the short $osp(2,2)$ IRR's as well.

The dimension of the Hilbert space with $N = n$ is $2n + 1$. We denote it by $\mathcal{H}^{2n+1}$. It is the direct sum $\mathcal{H}^{n+1} \oplus \mathcal{H}^n$ where $\mathcal{H}^{n+1}$ is the even subspace carrying the IRR $\Gamma^0_{n/2}$ and $\mathcal{H}^n$ is the odd subspace carrying the representation $\Gamma^0_{(n-1)/2}$. A basis for $\mathcal{H}^{2n+1}$ is

$$\frac{(a^\dagger_1)^{n_1}}{\sqrt{n_1!}} \frac{(a^\dagger_2)^{n_2}}{\sqrt{n_2!}} (b^\dagger)^{n_3} |0\rangle \ , \quad \sum n_i = n \ , \quad n_3 \in (0,1) \ , \quad (b^\dagger)^0 := 1 \ . \tag{10.30}$$

The fuzzy SUSY $S_F^{2,2}$ (in the zero instanton sector) is the matrix algebra $Mat(4J+1) = Mat(2n+1)$. Just as S_F^2, it can be realized using oscillators. In terms of oscillators, a typical element is

$$\sum_{i,j} c_{i,j}^m (a_i^\dagger)^m (a_j)^m + \sum_{i,j} d_{i,j}^{m-1} (a_i^\dagger)^{m-1} (a_j)^{m-1} b^\dagger b , \quad c_{i,j}^m, d_{i,j}^{m-1} \in \mathbb{C} .$$

(10.31)

It is to be restricted to the space $\mathcal{H}^{2n+1}$.

The left- and right-actions

$$\Lambda_\rho^L M = \Lambda_\rho M , \quad \Lambda_\rho^R M = (-1)^{|\Lambda_\rho||M|} M \Lambda_\rho$$

(10.32)

of $osp(2,\mathcal{N})$ on $Mat(2n+1)$ give two commuting IRR's of $osp(2,\mathcal{N})$. Here, $\Lambda_\rho \in osp(2,\mathcal{N})$, $\mathcal{N} = 1,2$, $M \in Mat(2n+1)$ and both Λ_ρ and M are of definite grade $|\Lambda_\rho|, |M|$ (mod 2).

Combining the left- and right- representations, we get the grade adjoint representation

$$\text{gad} : \Lambda_\rho \to \text{gad}\Lambda_\rho = \Lambda_\rho^L - \Lambda_\rho^R , \quad \rho \in (i, \alpha, \alpha', 8)$$

(10.33)

of $osp(2,\mathcal{N})$.

With regard to gad, $Mat(4J+1)$ transforms as

$$\Gamma_J^1 \otimes \Gamma_J^1 = \Gamma_{2J}^1 \oplus \Gamma_{2J-1/2}^1 \oplus \Gamma_{2J-1}^1 \oplus \cdots \oplus \Gamma_0^1 .$$

(10.34)

$osp(2,2)$ acts on $Mat(4J+1)$ by L, R and gad representations as well. The L and R are the short representations $\Gamma_{J\pm}^2$ so that under gad, $Mat(4J+1)$ transforms as $\Gamma_{J+}^2 \otimes \Gamma_{J-}^2$. Its reduction can be inferred from (10.34) once we know that $\Gamma_J^2(0)|_{osp(2,1)} = \Gamma_J^1 \oplus \Gamma_{J-1/2}^1$. We will see this later. Hence

$$\Gamma_{J+}^2 \otimes \Gamma_{J-}^2 = \Gamma_{2J}^2(0) \oplus \Gamma_{2J-1}^2(0) \oplus \cdots \oplus \Gamma_0^2(0) .$$

(10.35)

The fuzzy field Φ is an element of fuzzy SUSY. The free action for Φ is

$$S_0 = \frac{f^2}{2} \, str \; \Phi^\ddagger V_0 \Phi ,$$

(10.36)

where f is a real constant and V_0 is an $osp(2,1)$-invariant operator. When restricted to the odd subspace, it should become the Dirac operator of chapter 8.

The limit of this operator for "$J = \infty$" (that is on $S^{2,2}$) was found by Fronsdal [127] and later used effectively by Grosse *et al* [7]. For $J = \infty$, it is the difference $K_1 - K_2$ of the Casimir operators K_1 and K_2 written as graded differential operators. This operator, for finite J, becomes

$$V_0 = \varepsilon_{\alpha\beta}(\Lambda_{\alpha'})(\Lambda_{\beta'}) + \frac{1}{4}(\Lambda_8)^2 .$$

(10.37)

It is evident that V_0 is $osp(2,1)$-invariant. But it is less obvious that gad $\Lambda_{\alpha'}$, gad Λ_8 *anti*-commute with V_0:

$$\{\text{gad}\Lambda_{\alpha'}, V_0\} = \{\text{gad}\Lambda_8, V_0\} = 0 \ . \tag{10.38}$$

This means that these generators are realized as *chiral* symmetries. Of these, gad Λ_8, restricted to the odd sector, is just standard chirality. Thus, these generators, associated with $osp(2,2)/osp(2,1)$ are SUSY generalizations of conventional chirality.

We now show these results.

10.2 SUSY Chirality

Let us first exhibit the highest weight vectors of the $su(2)$ IRR's which occur in $\Gamma_j^2(0)$. Here j is an integer. Referring to (10.12), we have that $\Gamma_j^1\big|_{su(2)} = \Gamma_j^0 \oplus \Gamma_{j-1/2}^0$ for $j \geq 1$. The highest weight vector of Γ_j^0 is $(a_1^\dagger a_2)^j$ as it commutes with Λ_4 and carries the eigenvalue j of gad Λ_3. gad Λ_5 maps it to $-j(a_1^\dagger a_2)^{j-1}\Lambda_4$, the highest weight vector of $\Gamma_{j-1/2}^0 \subset \Gamma_j^1$. Thus

$$
\begin{array}{ccc}
\Gamma_j^1\big|_{su(2)} = & \Gamma_j^0 & \oplus & \Gamma_{j-1/2}^0 \\
\text{Highest weight} \Big\} & (a_1^\dagger a_2)^j & \xrightarrow{\text{gad}\Lambda_5} & -j(a_1^\dagger a_2)^{j-1}\Lambda_4 \ , \quad j \geq 1 \ . \\
\text{vectors} & & &
\end{array}
\tag{10.39}
$$

The equation also indicates the operator mapping one highest weight vector of $su(2)$ to the other.

Next consider $\Gamma_{j-1/2}^1 \supset \Gamma_{j-1/2}^0 \oplus \Gamma_{j-1}^0$ for $j \geq 1$. To distinguish the $su(2)$ IRR's here from those in Γ_j^1, we put a prime on them:

$$\Gamma_{j-1/2}^1\Big|_{su(2)} = \Gamma_{j-1/2}^{0'} \oplus \Gamma_{j-1}^{0'} \ . \tag{10.40}$$

The highest weight state of $\Gamma_{j-1/2}^1$, commuting with Λ_4 and with eigenvalue $j-1/2$ for gad Λ_3 is $-j(a_1^\dagger a_2)^{j-1}\Lambda_6$. And Λ_5 maps it to the highest weight vector X_{j-1} of $\Gamma_{j-1}^{0'}$. We show X_{j-1} below. Thus

$$
\begin{array}{ccc}
\Gamma_{j-1/2}^1\Big|_{su(2)} = & \Gamma_{j-1/2}^{0'} & \oplus & \Gamma_{j-1}^{0'} , \\
\text{Highest weight} \Big\} & -j(a_1^\dagger a_2)^{j-1}\Lambda_6 & \xrightarrow{\text{gad}\Lambda_5} & X_{j-1} , \\
\text{vectors} & & &
\end{array}
\tag{10.41}
$$

$$X_{j-1} = \frac{j-2J-1}{4}(a_1^\dagger a_2)^{j-1} + \frac{1-2j}{4}(a_1^\dagger a_2)^{j-1}b^\dagger b \ , \quad j \geq 1 \ .$$

In calculating X_{j-1}, we use

$$\Lambda_4\Lambda_6 = -\frac{1}{4}(a_1^\dagger a_2)(2b^\dagger b - 1) \ , \quad a^\dagger a + b^\dagger b = 2J \ . \tag{10.42}$$

Now gad Λ_7, gad Λ_8 map the vectors in (10.39) to the vectors in (10.41). The full table is

$$
\begin{array}{ccc}
\Gamma_j^1 \ni & (a_1^\dagger a_2)^j & \xrightarrow{\;\text{gad}\Lambda_5\;} & -j(a_1^\dagger a_2)^{j-1}\Lambda_4 \\[2pt]
\nearrow & \in \Gamma_j^0 & & \in \Gamma_{j-1/2}^0 \\
\Gamma_j^2(0) & \text{gad}\Lambda_7 \downarrow & \nearrow\, \text{gad}\Lambda_8 & \downarrow \text{gad}\Lambda_7 \\
\searrow & & & \\[4pt]
& \Gamma_{j-1/2}^1 \ni -j(a_1^\dagger a_2)^{j-1}\Lambda_6 & \xrightarrow{\;\text{gad}\Lambda_5\;} & X_{j-1} \quad , \qquad j \geq 1 \;. \\[2pt]
& \in \Gamma_{j-1/2}^{0'} & & \in \Gamma_{j-1}^{0'}
\end{array}
$$

$$(10.43)$$

For $j = 0$, we get the trivial IRR of $osp(2,\mathcal{N})$'s.

Eq. (10.43) shows that gad $\Lambda_{\alpha'}$, gad Λ_8 map the vectors of Γ_j^1 to those of $\Gamma_{j-1/2}^1$ ($j \geq 1$) and vice versa. So if V_0 has opposite eigenvalues in the representations in (10.43), then we can conclude that[*]

$$\{\text{gad}\Lambda_{\alpha'}, V_0\} = \{\text{gad}\Lambda_8, V_0\} = 0 \tag{10.44}$$

identically, since $V_0|_{\Gamma_0^1} = 0$. That means that these operators associated with $osp(2,2)/osp(2,1)$ are *chirally* realized symmetries.

10.3 Eigenvalues of V_0

As V_0 is an $osp(2,1)$ scalar, it is enough to compute its eigenvalue on the highest weight state of Γ_j^1 to find $V_0|_{\Gamma_j^1}$.

As $\Lambda_{4'} = \Lambda_6$ commutes with $(a_1^\dagger a_2)^j$, we have that

$$\varepsilon_{\alpha\beta}\, \text{gad}\Lambda_{\alpha'}\, \text{gad}\Lambda_{\beta'}\, (a_1^\dagger a_2)^j = (\, \text{gad}\Lambda_{4'}\, \text{gad}\Lambda_{5'} + \text{gad}\Lambda_{5'}\, \text{gad}\Lambda_{4'}\,)\, (a_1^\dagger a_2)^j \tag{10.45}$$

where the sign of the second term has been switched as it is zero anyway. Thus the left-hand side of the previous formula can be written as

$$\text{gad}\{\Lambda_{4'}, \Lambda_{5'}\}(a_1^\dagger a_2)^j = \frac{1}{2}\text{gad}\Lambda_3(a_1^\dagger a_2)^j = \frac{j}{2}(a_1^\dagger a_2)^j \;. \tag{10.46}$$

Also

$$\text{gad}\Lambda_8(a_1^\dagger a_2)^j = 0 \;. \tag{10.47}$$

Hence

$$V_0(a_1^\dagger a_2)^j = \frac{j}{2}(a_1^\dagger a_2)^j \;. \tag{10.48}$$

---

[*]To show that $\{\text{gad}\Lambda_6, V_0\} = 0$ we use the fact that $\text{gad}\Lambda_6 = -[\text{gad}\Lambda_4, \text{gad}\Lambda_8]$. The result follows from the graded Jacobi identity.

One quick way to evaluate $V_0|_{\Gamma^1_{j-1/2}}$ is as follows. Since $K_1|_{\Gamma^1_j} = j(j + 1/2)$, we have

$$K_2|_{\Gamma^1_j} = (K_1 - V_0)|_{\Gamma^1_j} = j^2 . \tag{10.49}$$

But K_2 is $osp(2,2)$-invariant. Hence

$$K_2|_{\Gamma^1_{j-1/2}} = j^2 . \tag{10.50}$$

Since also $K_1|_{\Gamma^1_{j-1/2}} = j(j - 1/2)$, we have

$$V_0|_{\Gamma^1_{j-1/2}} = (K_1 - K_2)|_{\Gamma^1_{j-1/2}} = -\frac{j}{2}\mathbb{1} . \tag{10.51}$$

Thus V_0 has opposite eigenvalues on Γ^1_j and $\Gamma^1_{j-1/2}$.

It is important to notice that

$$K_2 = (2V_0)^2 . \tag{10.52}$$

That is, $2V_0$ is a square root of K_2, a little like in the way that the Dirac operator is the square root of the Laplacian.

10.4 Fuzzy SUSY Instantons

The manifold S^2 admits twisted $U(1)$ bundles labelled by a topological index or Chern number $k \in \mathbb{Z}$. In the algebraic language, sections of vector bundles associated with these $U(1)$ bundles are described by elements of projective modules (see chapter 5,6).

When S^2 becomes the graded supersphere $S^{2,2}$, we expect these modules to persist, and become in some sense supersymmetric projective modules. That is in fact the case. We shall see that explicitly after first studying their fuzzy analogues.

The projective modules on S^2 and S^2_F are associated with $SU(2) \simeq S^3$ via Hopf fibration and Lens spaces. In the same way, the supersymmetric projective modules on $S^{2,2}$ and $S^{2,2}_F$ get associated with $osp(2,1)$ and $osp(2,2)$.

The fuzzy algebra $S^{2,2}_F$ of previous sections is to be assigned $k = 0$. The elements of this algebra are square matrices mapping the space with $N = 2J$ to the same space $N = 2J$. We emphasize the value of k by writing $S^{2,2}_F(0)$ for $S^{2,2}_F$. $S^{2,2}_F(0)$ is a bimodule for $osp(2,2)$ as the latter can act on the left or right of $S^{2,2}_F(0)$ by the IRR's $\Gamma^2_{J\pm}(0)$.

For $k \neq 0$, $S^{2,2}_F(k)$ is not an algebra. It can be described using projectors or equally well as maps of the vector space with $N = 2J$ to the one with

$N = 2J + k$. (We take $J + \frac{k}{2} \geq 0$. If $k < 0$, this means $J \geq \frac{|k|}{2}$.) If a basis is chosen for the domain and range of $S_F^{2,2}(k)$, their elements become rectangular matrices with $2J + k$ rows and $2J$ columns. $S_F^{2,2}(k)$ as well is a bimodule for $osp(2,2)$. The latter acts by $\Gamma^2_{(J+\frac{k}{2})+}$ on the left of $S_F^{2,2}(k)$ and by Γ^2_{J-} on the right of $S_F^{2,2}(k)$.

The invariant associated with $S_F^{2,2}(k)$ is just k. The meaning of k is

$$k = \text{Dimension of range of } S_F^{2,2}(k) - \text{Dimension of domain of } S_F^{2,2}(k) \,.$$
(10.53)

Scalar fields Φ are now elements of $S_F^{2,2}(k)$ while V_0 is replaced by a new operator V_k which incorporates the appropriate connection and "topological" data. We now argue, using index theory and other considerations, that the $osp(2,1)$-invariant V_k is fixed by the requirement

$$V_k^2 = K_2$$
(10.54)

where K_2 is the Casimir invariant for $\Gamma^2_{(J+\frac{k}{2})+} \otimes \Gamma^2_{J-}$.

10.5 Fuzzy SUSY Zero Modes and their Index Theory

We begin by analyzing the $osp(2,1)$ and $osp(2,2)$ representation content of $S_F^{2,2}(k)$.

As regards the gad representation of $osp(2,1)$, it transforms according to

$$\Gamma^1_{J+\frac{k}{2}} \otimes \Gamma^1_J = \left(\Gamma^1_{2J+\frac{k}{2}} \oplus \Gamma^1_{2J+\frac{k}{2}-\frac{1}{2}} \right)$$
$$\oplus \left(\Gamma^1_{2J+\frac{k}{2}-1} \oplus \Gamma^1_{2J+\frac{k}{2}-\frac{3}{2}} \right) \oplus \cdots \oplus \left(\Gamma^1_{\frac{|k|}{2}+1} \oplus \Gamma^1_{\frac{|k|}{2}+\frac{1}{2}} \right) \oplus \Gamma^1_{\frac{|k|}{2}} \,.$$

The analogue of (10.43) is:

$$2J + \frac{k}{2} \geq j \geq \frac{|k|}{2} + 1 \,,$$

$$\Gamma^1_j \ \ni \ \Gamma^0_j \ \oplus \ \Gamma^0_{j-1/2}$$

$$\Gamma^2_j(k) \nearrow$$
(10.55)

$$\searrow \Gamma^1_{j-1/2} \ni \Gamma^0_{j-1/2} \oplus \Gamma^0_{j-1} \quad .$$

Here $|k| \geq 1$. For $j = \frac{|k|}{2}$, we get the atypical representation of $osp(2,2)$:

$$\Gamma^2_{\frac{|k|}{2}}(k) \to \Gamma^1_{\frac{|k|}{2}} = \Gamma^0_{\frac{|k|}{2}} \oplus \Gamma^0_{\frac{|k|}{2}-1} \,.$$
(10.56)

All this becomes explicit during the following calculation of the eigenvalues of K_2.

10.5.1 Spectrum of K_2

For $k > 0$, the highest weight vector with angular momentum

$$j = m + \frac{|k|}{2} , \quad m = 0, 1, \dots \tag{10.57}$$

is

$$(a_1^\dagger)^{|k|} (a_1^\dagger a_2)^m . \tag{10.58}$$

Since

$$\mathrm{gad}\Lambda_8 (a_1^\dagger)^{|k|} (a_1^\dagger a_2)^m = |k| (a_1^\dagger)^{|k|} (a_1^\dagger a_2)^m , \tag{10.59}$$

it is the highest weight vector of $\Gamma_j^2(|k|)$. Thus $\Gamma_j^2(|k|)$ occurs in the reduction of the $osp(2,2)$ action on $S_F^{2,2}(|k|)$.

General theory [126] tells us the branching rules of $\Gamma_j^2(|k|)$ as in (10.55). This equation is thus established for $k > 0$.

We can check as before that

$$\varepsilon_{\alpha\beta} \, \mathrm{gad}\Lambda_{\alpha'} \, \mathrm{gad}\Lambda_{\beta'} \, (a_1^\dagger)^{|k|} (a_1^\dagger a_2)^m = \frac{1}{2} \left(m + \frac{|k|}{2} \right) (a_1^\dagger)^{|k|} (a_1^\dagger a_2)^m \tag{10.60}$$

while

$$\frac{1}{4} (\mathrm{gad}\Lambda_8)^2 (a_1^\dagger)^{|k|} (a_1^\dagger a_2)^m = \frac{k^2}{4} \tag{10.61}$$

and

$$K_1 |_{\Gamma_j^1} = j(j+1) \mathbb{1} . \tag{10.62}$$

We thus have [126]

$$K_2 |_{\Gamma_j^2(|k|)} = \left(j^2 - \frac{k^2}{4} \right) \mathbb{1} . \tag{10.63}$$

For $k < 0$,

$$(a_2)^{|k|} (a_1^\dagger a_2)^m \tag{10.64}$$

is the highest weight vector for angular momentum

$$j = m + \frac{|k|}{2} . \tag{10.65}$$

Since

$$\mathrm{gad}\Lambda_8 (a_2)^{|k|} (a_1^\dagger a_2)^m = -|k| (a_2)^{|k|} (a_1^\dagger a_2)^m , \tag{10.66}$$

it is the highest weight vector of $\Gamma_j^2(-|k|)$. Hence $\Gamma_j^2(-|k|)$ occurs in the reduction of the $osp(2,2)$ action on $S_F^{2,2}(-|k|)$. We thus establish (10.55) for $k < 0$ as well.

The eigenvalues of V_k, when restricted to Γ_j^1 and $\Gamma_{j-1/2}^1$ and for $j \geq \frac{|k|}{2} + 1$, are $\pm\sqrt{j^2 - \frac{k^2}{4}}$. These eigenvalues are not zero. Hence the $osp(2,2)$ operators which intertwine these representations, mapping vectors of one representation to the other, *anticommute* with V_k: they are *chiral* symmetries for these representations. For $j = \frac{|k|}{2}$, V_k vanishes while the representation space carries the atypical representation $\Gamma_{\frac{|k|}{2}}^2(k)$ of $osp(2,2)$. Thus we can say that the above chiral operators all anticommute with $V_k|_{J=0}$. Hence these operators anticommute with V_k (for any j, on all vectors of $S_F^{2,2}(k)$) just as standard chirality anticommutes with the massless Dirac operator.

For $k = 0$, these operators were $\Lambda_{\alpha'}$, Λ_8. But they change with k. They can be worked out. They do not occur in subsequent discussion and hence we do not show them here.

We now establish that V_k is the correct choice for the fuzzy SUSY action S_k:

$$S_k = \text{const } str\, \Phi^\dagger V_k \, \Phi \ . \tag{10.67}$$

This formula is valid also for $k = 0$ as we saw earlier. We here focus on $k \neq 0$.

The Dirac operators D for fuzzy spheres of instanton number k are known (see chapter 8). We first show that V_k coincides with this operator on the Dirac sector.

It is enough to focus on typical $osp(2,2)$ IRR's since both the Dirac operator and V_k vanish on the grade-odd sector of $\Gamma_{\frac{|k|}{2}}^1$. Thus consider $\Gamma_j^2(k)$ for $j \geq \frac{1}{2}|k| + 1$. Angular momentum J in the Dirac sector of $\Gamma_j^2(k)$ is $j - 1/2$. Hence

$$V_k^2\big|_{\Gamma_j^2(k)\ Dirac\ sector} = \left(J - \frac{|k| - 1}{2}\right)\left(J + \frac{|k| + 1}{2}\right) \mathbb{1} \ . \tag{10.68}$$

Substituting $J = n + \frac{|k|-1}{2}$ and identifying $|k| = 2T$, we get the answer:

$$V_k^2\big|_{\Gamma_j^2(k)\ Dirac\ sector} = n(n + 2T)\mathbb{1} \ . \tag{10.69}$$

Hence $V_k^2\big|_{\Gamma_j^2(k)\ Dirac\ sector}$ is the correct Dirac operator .

This result and the $osp(2,1)$-invariance of V_k are compelling reasons to identify it as the SUSY generalization of the Dirac and Laplacian operators for $k \neq 0$.

10.5.2 Index Theory and Zero Modes

There is also further evidence supporting the correctness of V_k: It gives the SUSY generalization of index theory.

Thus one knows from chapter 8 that 1) the Dirac operator has $|k|$ zero modes for instanton number k on S^2 and on the fuzzy sphere $S_F^2(k)$, and that 2) they are left- (right-) chiral if $k > 0$ ($k < 0$), 3) charge conjugation interchanges these chiralities.

More precisely if $n_{L,R}$ are the number of left- and right-chiral zero modes,

$$n_L - n_R = k \ . \tag{10.70}$$

This number is "topologically stable". The meaning of this statement in the fuzzy case can be found in chapter 8.

If the Dirac operator is $SU(2)$-invariant, these zero modes organize themselves into $SU(2)$ multiplets with angular momentum $\frac{|k|}{2}$.

Now V_k has zero modes which form the atypical multiplet $\Gamma^2_{\frac{|k|}{2}}(k)$ of $osp(2,2)$. The number of zero modes is $2|k| + 1$. Of these, $|k|$ correspond to the grade odd sector and can be identified with the zero modes of S^2 and $S_F^2(k)$ Dirac operators. The remaining grade even ($|k| + 1$) zero modes are their SUSY-partners.

The zero modes transform by inequivalent IRR's of $osp(2,2)$ for the two signs of k. These two atypical $osp(2,2)$ representations are SUSY generalizations of left- and right- chiralities.

Identifying charge conjugation with the automorphism (10.19), we see that it exchanges these two IRR's just as it exchanges chiralities in the Dirac sector.

10.6   Final Remarks

In this chapter, we have extended the developments of chapter 9 on the fuzzy SUSY model on S^2 to the instanton sector. A SUSY generalization of chapter 8 on index theory and zero modes of the Dirac operator has also been established.

We can try introducing interactions involving just Φ. For $k \neq 0$, Φ can be thought of as a rectangular matrix. So $\Phi^\ddagger \Phi$ and $\Phi \Phi^\ddagger$ are square matrices of different sizes acting on $osp(2,2)$ representations with $N = n$ and $N = n + k$. A typical interaction may then be

$$\lambda str(\Phi^\ddagger \Phi)^2 \tag{10.71}$$

where *str* is over the space with $N = n$, the domain of $\Phi^\ddagger\Phi$. But note that (10.71) and the use of *str* in interactions require further study.

Fuzzy SUSY gauge theories remain to be formulated. The investigation of the graded commutative limit $n \to \infty$ with k fixed has also not been done for $k \neq 0$.

Chapter 11

Fuzzy Spaces as Hopf Algebras

11.1 Overview

In this chapter we will explore yet another intriguing aspect of fuzzy spaces, namely their potential use as quantum symmetry algebras. To be more precise, we will establish, through studying the fuzzy sphere as an example, that fuzzy spaces possess a Hopf algebra structure.

It is a fact that for an algebra $\mathcal{A}$, it is not always possible to compose two of its representations ρ and σ to obtain a third one. For groups we can do so and obtain the tensor product $\rho \otimes \sigma$. Such a composition of representations is also possible for coalgebras $\mathcal{C}$ [128]. A coalgebra $\mathcal{C}$ has a coproduct Δ which is a homomorphism from $\mathcal{C}$ to $\mathcal{C} \otimes \mathcal{C}$ and the composition of its representations ρ and σ is the map $(\rho \otimes \sigma)\Delta$. If $\mathcal{C}$ has a more refined structure and is a Hopf algebra, then it closely resembles a group, in fact sufficiently so that it can be used as a "quantum symmetry group" [129].

We follow reference [130] in this chapter. In order to make our discussion self contained, we review some of the basic definitions about coalgebras, bialgebras and Hopf algebras in terms of the language of commutative diagrams and set our notations and conventions, which are the standard ones in the literature. A well known example of a Hopf algebra is the group algebra G^* associated to a group G. Our interest mainly lies on compact Lie groups G, as they are the ones whose adjoint orbits once quantized yield fuzzy spaces. The group algebra G^* of such G consists of elements $\int_G d\mu(g)\alpha(g)g$ where $\alpha(g)$ is a smooth complex function and $d\mu(g)$ is the G-invariant measure. It is isomorphic to the convolution algebra of functions on G. Basic definitions and properties related to G^* will be given in section 11.3.

In section 11.5 and 11.6, we establish that fuzzy spaces are irreducible representations ρ of G^* and inherit its Hopf algebra structure. For fixed G, their direct sum is homomorphic to G^*. For example both $S_F^2(J)$ and $\oplus_J S_F^2(J) \simeq SU(2)^*$ are Hopf algebras. This means that we can define a co-product on $S_F^2(J)$ and $\oplus_J S_F^2(J)$ and compose two fuzzy spheres preserving algebraic properties intact.

A group algebra G^* and a fuzzy space from a group G carry several actions of G. G acts on G and G^* by left and right multiplications and by conjugation. Also for example, the fuzzy space $S_F^2(J)$ consists of $(2J + 1) \times (2J + 1)$ matrices and the spin J representation of $SU(2)$ acts on these matrices by left and right multiplication and by conjugation. The map ρ of G^* to a fuzzy space and the coproduct Δ are compatible with all these actions: they are G-equivariant.

Elements m of fuzzy spaces being matrices, we can take their hermitian conjugates. They are *-algebras if * is hermitian conjugation. G^* also is a *-algebra. ρ and Δ are *-homomorphisms as well: $\rho(\alpha^*) = \rho(\alpha)^\dagger$, $\Delta(m^*) = \Delta(m)^*$.

The last two properties of Δ on fuzzy spaces also derive from the same properties of Δ for G^*.

All this means that fuzzy spaces can be used as symmetry algebras. In that context however, G-invariance implies G^*- invariance and we can substitute the familiar group invariance for fuzzy space invariance.

The remarkable significance of the Hopf structure seems to lie elsewhere. Fuzzy spaces approximate space-time algebras. $S_F^2(J)$ is an approximation to the Euclidean version of (causal) de Sitter space homeomorphic to $S^1 \times \mathbb{R}$, or for large radii of S^1, of Minkowski space [131]. The Hopf structure then gives orderly rules for splitting and joining fuzzy spaces. The decomposition of $(\rho \otimes \sigma)\Delta$ into irreducible *-representations (IRR's) τ gives fusion rules for states in ρ and σ combining to become τ, while Δ on an IRR such as τ gives amplitudes for τ becoming ρ and σ. In other words, Δ gives Clebsch-Gordan coefficients for space-times joining and splitting. Equivariance means that these processes occur compatibly with G-invariance: G gives selection rules for these processes in the ordinary sense. The Hopf structure has a further remarkable consequence: An observable on a state in τ can be split into observables on its decay products in ρ and σ.

There are similar results for field theories on τ, ρ and σ, indicating the possibility of many orderly calculations.

These mathematical results are very suggestive, but their physical consequences are yet to be explored.

The coproduct Δ on the matrix algebra $Mat(N + 1)$ is not unique. Its choice depends on the group actions we care to preserve, that of $SU(2)$ for S_F^2, that of $SU(N+1)$ for the fuzzy $\mathbb{C}P^N$ algebra $\mathbb{C}P_F^N$ and so forth. It is thus the particular equivariance that determines the choice of Δ.

We focus attention on the fuzzy sphere for specificity in what follows, but one can see that the arguments are valid for any fuzzy space. Proofs for the fuzzy sphere are thus often assumed to be valid for any fuzzy space without comment.

Fuzzy algebras such as $\mathbb{C}P_F^N$ can be further "q-deformed" into certain quantum group algebras relevant for the study of D-branes. This theory has been developed in detail by Pawelczyk and Steinacker [132].

11.2 Basics

Here we collect some of the basic formulae related to the group $SU(2)$ and its representations which will be used later in the chapter.

The canonical angular momentum generators of $SU(2)$ are J_i ($i = 1, 2, 3$). The unitary irreducible representations (UIRR's) of $SU(2)$ act for any half-integer or integer J on Hilbert spaces $\mathcal{H}^J$ of dimension $2J+1$. They have orthonormal basis $|J, M\rangle$, with $J_3|J, M\rangle = M|J, M\rangle$ and obeying conventional phase conventions. The unitary matrix $D^J(g)$ of $g \in SU(2)$ acting on $\mathcal{H}^J$ has matrix elements $\langle J, M|D^J(g)|J, N\rangle = D^J(g)_{MN}$ in this basis.

Let

$$V = \int_{SU(2)} d\mu(g) \tag{11.1}$$

be the volume of $SU(2)$ with respect to the Haar measure $d\mu$. It is then well-known that [133]

$$\int_{SU(2)} d\mu(g) D^J(g)_{ij} \, D^K(g)^\dagger_{kl} = \frac{V}{2J+1} \, \delta_{JK} \, \delta_{il} \, \delta_{jk} \,, \tag{11.2a}$$

$$\frac{2J+1}{V} \sum_{J,ij} D^J_{ij}(g) \, \bar{D}^J_{ij}(g') = \delta_g(g') \,, \tag{11.2b}$$

where bar stands for complex conjugation and δ_g is the δ-function on $SU(2)$ supported at g:

$$\int_{SU(2)} d\mu(g') \, \delta_g(g') \alpha(g') = \alpha(g) \tag{11.3}$$

for smooth functions α on G.

We have also the Clebsch-Gordan series

$$D^K_{\mu_1 m_1} D^L_{\mu_2 m_2} = \sum_J C(K,L,J\,;\mu_1\,,\mu_2)\,C(K,L,J\,;m_1\,,m_2)\,D^J_{\mu_1+\mu_2\,,m_1+m_2}$$

(11.4)

where C's are the Clebsch-Gordan coefficients.

11.3   The Group and the Convolution Algebras

The group algebra consists of the linear combinations

$$\int_G d\mu(g)\,\alpha(g)\,g\,, \qquad d\mu(g) = \text{Haar measure on } G \qquad (11.5)$$

of elements g of G, α being any smooth $\mathbb{C}$-valued function on G. The algebra product is induced from the group product:

$$\int_G d\mu(g)\,\alpha(g)\,g \int_G d\mu(g')\,\beta(g')\,g' := \int_G d\mu(g) \int_G d\mu(g')\alpha(g)\,\beta(g')(gg')\,.$$

(11.6)

We will henceforth omit the symbol G under integrals.

The right hand side of (11.6) is

$$\int d\mu(s)\,(\alpha *_c \beta)(s)\,s \qquad (11.7)$$

where $*_c$ is the convolution product:

$$(\alpha *_c \beta)(s) = \int d\mu(g)\alpha(g)\,\beta(g^{-1}s)\,. \qquad (11.8)$$

The convolution algebra consists of smooth functions α on G with $*_c$ as their product. Under the map

$$\int d\mu(g)\alpha(g)g \to \alpha\,, \qquad (11.9)$$

(11.6) goes over to $\alpha *_c \beta$ so that the group algebra and convolution algebra are isomorphic. We call either as G^*.

Using invariance properties of $d\mu$, (11.9) shows that under the action

$$\int d\mu(g)\,\alpha(g)\,g \to h_1 \left(\int d\mu(g)\alpha(g)g \right) h_2^{-1}$$

$$= \int d\mu(g)\alpha(g)h_1 g h_2^{-1}\,, \qquad h_i \in G, \quad (11.10)$$

$\alpha \to \alpha'$ where

$$\alpha'(g) = \alpha(h_1^{-1}g h_2)\,. \qquad (11.11)$$

Thus the map (11.9) is compatible with left- and right- G-actions.

The group algebra is a $*$-algebra [128], the $*$-operation being

$$\left[\int d\mu(g)\,\alpha(g)\,g \right]^* = \int d\mu(g)\,\bar{\alpha}(g)g^{-1}\,. \tag{11.12}$$

The $*$-operation in G^* is

$$* : \alpha \to \alpha^*\,,$$
$$\alpha^*(g) = \bar{\alpha}(g^{-1})\,. \tag{11.13}$$

Under the map (11.9),

$$\left[\int d\mu(g)\alpha(g)g \right]^* \to \alpha^* \tag{11.14}$$

since

$$d\mu(g) = i\,Tr(g^{-1}\,dg) \wedge g^{-1}\,dg \wedge g^{-1}\,dg = -d\mu(g^{-1})\,. \tag{11.15}$$

The minus sign in (11.15) is compensated by flips in "limits of integration", thus $\int d\mu(g) = \int d\mu(g^{-1}) = V$. Hence the map (11.9) is a $*$-morphism, that is, it preserves "hermitian conjugation".

11.4   A Prelude to Hopf Algebras

This section reviews the basic ingredients that go into the definition of Hopf algebras. It also sets some notations and conventions which are standard in the literature. Our approach here will be illustrative and will closely follow the exposition of [135]. Unless stated otherwise, we always work over the complex number field $\mathbb{C}$, but the definitions given below extend to any number field k without the need for further remarks.

In the language of commutative diagrams, an algebra $\mathcal{A}$ is defined as the triple $\mathcal{A} \equiv (\mathcal{A}, M, u)$ where $\mathcal{A}$ is a vector space, $M : \mathcal{A} \otimes \mathcal{A} \to \mathcal{A}$ and $u : \mathbb{C} \to \mathcal{A}$ are morphisms (linear maps) of vector spaces such that the following diagrams are commutative.

$$
\begin{array}{ccc}
\mathcal{A} \otimes \mathcal{A} \otimes \mathcal{A} & \xrightarrow{\;id \otimes M\;} & \mathcal{A} \otimes \mathcal{A} \\
{\scriptstyle M \otimes id}\Big\downarrow & & \Big\downarrow{\scriptstyle M} \\
\mathcal{A} \otimes \mathcal{A} & \xrightarrow{\;\;\;M\;\;\;} & \mathcal{A}
\end{array}
$$

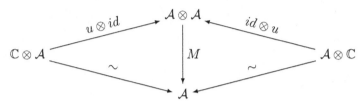

In this definition M is called the product and u is called the unit. The commutativity of the first diagram simply implies the associativity of the product M, whereas for the latter it expresses the fact that u is the unit of the algebra. The arrows labeled by $\sim$ are the canonical isomorphisms of the algebra onto itself. Also in above and what follows id denotes the identity map.

A coalgebra $\mathcal{C}$ is the triple $\mathcal{C} \equiv (\mathcal{C}, \Delta, \varepsilon)$, where $\mathcal{C}$ is a vector space, $\Delta : \mathcal{C} \to \mathcal{C} \otimes \mathcal{C}$ and $\varepsilon : \mathcal{C} \to \mathbb{C}$ are morphisms of vector spaces such that the following diagrams are commutative.

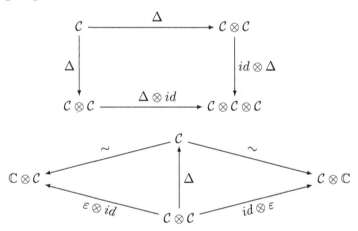

In this definition Δ is called the coproduct and ε is called the counit. The commutativity of the first diagram implies the coassociativity of the coproduct Δ, whereas for the latter it expresses the fact that ε is the counit of the coalgebra.

An immediate example of a coalgebra is the vector space of $n \times n$ matrices $Mat(n)$, with the coproduct and the counit

$$\Delta(e^{ij}) = \sum_{1 \leq p \leq n} e^{ip} \otimes e^{pj}, \quad \varepsilon(e^{ij}) = \delta^{ij},$$

$$(e^{ij})_{\alpha\beta} = \delta^i_\alpha \delta^j_\beta. \tag{11.16}$$

As e^{ij}, $1 \leq i, j \leq n$ is a basis for $Mat(n)^*$. Δ can be extended by linearity to all of C^*.

In what follows we adopt the "sigma" or "Sweedler" notation which is standard in literature and write for $c \in C$

$$\Delta(c) = \sum c_1 \otimes c_2, \qquad (11.17)$$

which with the usual summation convention should have been

$$\Delta(c) = \sum_{i=1}^{n} c_{i1} \otimes c_{i2}. \qquad (11.18)$$

One by one we are exhausting the steps leading to the definition of a Hopf algebra. The next step is to define the bialgebra structure. A bialgebra is a vector space H endowed with both an algebra and a coalgebra structure such that the following diagrams are commutative.

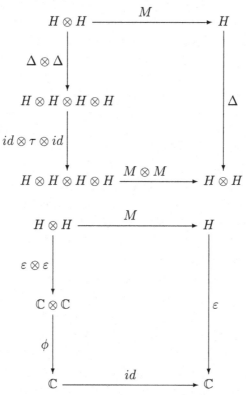

---

*One might be tempted to call (11.16) as the coproduct of S_F^2, since elements of S_F^2 are described by matrices in $Mat(n + 1)$. But, (11.16) is not equivariant under $SU(2)$ actions and therefore has no chance of being the appropriate coproduct for S_F^2.

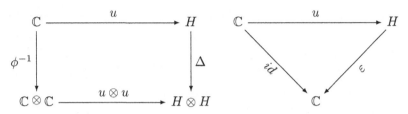

In the above, $\tau : H \otimes H \to H \otimes H$ is the twist map defined by $\tau(h_1 \otimes h_2) = h_2 \otimes h_1$, $\forall\, h_{1,2} \in H$. In terms of the sigma notation, the above four diagrams read

$$\Delta(hg) = \sum h_1 g_1 \otimes h_2 g_2 , \quad \varepsilon(hg) = \varepsilon(h)\varepsilon(g)$$
$$\Delta(1) = 1 \otimes 1 , \quad \varepsilon(1) = 1 . \tag{11.19}$$

Now, let S be a map from a bialgebra H onto itself. Then S is called an antipode if the following diagram is commutative.

$$
\begin{array}{ccccc}
H & \xrightarrow{\;\varepsilon\;} & \mathbb{C} & \xrightarrow{\;u\;} & H \\
{\scriptstyle\Delta}\big\downarrow & & & & \big\uparrow{\scriptstyle M} \\
H \otimes H & \xrightarrow{\quad id \otimes S\, ,\, S \otimes id \quad} & & & H \otimes H
\end{array}
$$

In terms of the sigma notation, this means

$$\sum S(h_1)h_2 = \sum h_1 S(h_2) = \varepsilon(h)\mathbf{1} , \qquad \mathbf{1} \in H . \tag{11.20}$$

By definition a *Hopf algebra* is a bialgebra with an antipode. Perhaps, the simplest example for a Hopf algebra is the group algebra, and it also happens to be the one of our interest. The group algebra G^* can be made into a Hopf algebra by defining the coproduct Δ, the counit ε and antipode S as follows:

$$\Delta(g) = g \otimes g , \tag{11.21a}$$
$$\varepsilon(g) = 1 \in \mathbb{C} , \tag{11.21b}$$
$$S(g) = g^{-1} . \tag{11.21c}$$

Here ε is the one-dimensional trivial representation of G and S maps g to its inverse. Δ, ε and S fulfill all the consistency conditions implied by the commutativity of the diagrams defining the Hopf algebra structure as can easily be verified. For instance we have

$$S(g)g = g^{-1}g = \mathbf{1} = \varepsilon(g)\mathbf{1} , \tag{11.22}$$

and similarly $gS(g) = \varepsilon(g)\mathbf{1}$ for any $g \in G$.

11.5   The *-Homomorphism $G^* \to S_F^2$

As mentioned earlier henceforth, we identify the group and convolution algebras and denote either by G^*. We specialize to $SU(2)$ for simplicity. We work with group algebra and group elements, but one may prefer the convolution algebra instead for reasons of rigor. (The image of g is the Dirac distribution δ_g and not a smooth function.)

The fuzzy sphere algebra is not unique, but depends on the angular momentum J as shown by the notation $S_F^2(J)$, which is $Mat(2J+1)$. Let

$$S_F^2 = \oplus_J S_F^2(J) = \oplus_J Mat(2J+1) \,. \tag{11.23}$$

Let $\rho(J)$ be the unitary irreducible representation of angular momentum J for $SU(2)$:

$$\rho(J): \quad g \to \langle \rho(J), g \rangle := D^J(g) \,. \tag{11.24}$$

We have

$$\langle \rho(J), g \rangle \, \langle \rho(J), h \rangle = \langle \rho(J), gh \rangle \,. \tag{11.25}$$

Choosing the *-operation on $D^J(g)$ as hermitian conjugation, $\rho(J)$ extends by linearity to a *-homomorphism on G^*:

$$\left\langle \rho(J), \int d\mu(g)\alpha(g)g \right\rangle = \int d\mu(g)\alpha(g)D^J(g)$$

$$\left\langle \rho(J), \left(\int d\mu(g)\alpha(g)g \right)^* \right\rangle = \int d\mu(g)\bar{\alpha}(g)D^J(g)^\dagger \,. \tag{11.26}$$

$\rho(J)$ is also compatible with group actions on G^* (that is, it is equivariant with respect to these actions):

$$\left\langle \rho(J), \int d\mu(g)\alpha(g)h_1 g h_2^{-1} \right\rangle =$$

$$\int d\mu(g)\alpha(g)D^J(h_1)D^J(g)D^J(h_2^{-1}) \,, \quad h_i \in SU(2) \,. \tag{11.27}$$

As by (11.2a),

$$\left\langle \rho(J), \frac{2K+1}{V} \int d\mu(g)(D_{ij}^K)^\dagger(g)g \right\rangle = e^{ji}(J)\delta_{KJ} \,,$$

$$e^{ji}(J)_{rs} = \delta_{jr}\delta_{is} \,, \quad i,j,r,s \in [-J, \cdots 0, \cdots, J] \,, \tag{11.28}$$

we see by (11.25) and (11.26) that $\rho(J)$ is a *-homomorphism from G^* to $S_F^2(J) \oplus \{0\}$, where $\{0\}$ denotes the zero elements of $\oplus_{K \neq J} S_F^2(K)$, the *-operation on $S_F^2(J)$ being hermitian conjugation. Identifying $S_F^2(J) \oplus \{0\}$ with $S_F^2(J)$, we thus get a *-homomorphism $\rho(J): G^* \to S_F^2(J)$. It is also seen to be equivariant with respect to $SU(2)$ actions, they are given on the basis $e^{ji}(J)$ by $D^J(h_1)e^{ji}(J)D^J(h_2)^{-1}$.

We can think of (11.26) as giving a map

$$\rho : g \quad \to \quad \langle \rho(.), g \rangle := g(.) \tag{11.29}$$

to a matrix valued function $g(.)$ on the space of UIRR's of $SU(2)$ where

$$g(J) = \langle \rho(J), g \rangle . \tag{11.30}$$

The homomorphism property (11.26) is expressed as the product $g(.)h(.)$ of these functions where

$$g(.)h(.)(J) = g(J)h(J) \tag{11.31}$$

is the point-wise product of matrices. This point of view is helpful for later discussions.

As emphasized earlier, this discussion works for any group G, its UIRR's, and its fuzzy spaces barring technical problems. Thus G^* is $*$-isomorphic to the $*$-algebra of functions $g(.)$ on the space of its UIRR's τ, with $g(\tau) = D^\tau(g)$, the linear operator of g in the UIRR τ and $g^*(\tau) = D^\tau(g)^\dagger$.

A fuzzy space is obtained by quantizing an adjoint orbit G/H, $H \subset G$ and approximates G/H. It is a full matrix algebra associated with a particular UIRR τ of G. There is thus a G-equivariant $*$-homomorphism from G^* to the fuzzy space.

At this point we encounter a difference with $S_F^2(J)$. For a given G/H we generally get only a subset of UIRR's τ. For example $\mathbb{C}P^2 = SU(3)/U(2)$ is associated with just the symmetric products of just 3's (or just 3*'s) of $SU(3)$. Thus the direct sum of matrix algebras from a given G/H is only homomorphic to G^*.

Henceforth we call the space of UIRR's of G as $\hat{G}$. For a compact group, $\hat{G}$ can be identified with the set of discrete parameters specifying all UIRR's.

The properties of a group G are captured by the algebra of matrix-valued functions $g(.)$ on $\hat{G}$ with point-wise multiplication, this algebra being isomorphic to G^*. In terms of $g(.)$, (11.21) translate to

$$\Delta(g(.)) = g(.) \otimes g(.) , \tag{11.32a}$$

$$\varepsilon(g(.)) = \mathbf{1} \in \mathbb{C} , \tag{11.32b}$$

$$S(g(.)) = g^{-1}(.) . \tag{11.32c}$$

Note that $g(.) \otimes g(.)$ is a function on $\hat{G} \otimes \hat{G}$.

11.6 Hopf Algebra for the Fuzzy Spaces

Any fuzzy space has a Hopf algebra structure, we show it here for the fuzzy sphere.

Let δ_J be the δ-function on $\widehat{SU(2)}$:

$$\delta_J(K) := \delta_{JK}.\tag{11.33}$$

(Since the sets of J and K are discrete, we have Kronecker delta and not a delta function).

Then

$$e^{ji}(J)\,\delta_J = \frac{2J+1}{V}\int d\mu(g)D_{ij}^J(g)^\dagger g(.)\tag{11.34}$$

Hence

$$\Delta(e^{ji}(J)\delta_J) = \frac{2J+1}{V}\int d\mu(g)D_{ij}^J(g)^\dagger g(.)\otimes g(.).\tag{11.35}$$

At $(K,L)\in\widehat{SU(2)}\otimes\widehat{SU(2)}$, this is

$$\Delta\big(e^{ji}(J)\big)(K,L) = \frac{2J+1}{V}\int d\mu(g)D_{ij}^J(g)^\dagger\,D^K(g)\otimes D^L(g).\tag{11.36}$$

As $\delta_J^2 = \delta_J$ and $\delta_J e^{ji}(J) = e^{ji}(J)\delta_J$, we can identify $e^{ji}(J)\delta_J$ with $e^{ji}(J)$:

$$e^{ji}(J)\delta_J \simeq e^{ji}(J).\tag{11.37}$$

Then (11.35) or (11.36) show that there are many coproducts $\Delta = \Delta_{KL}$ we can define and they are controlled by the choice of K and L:

$$\Delta\big(e^{ji}(J)\delta_J\big)(K,L) := \Delta_{KL}\big(e^{ji}(J)\big).\tag{11.38}$$

From section (11.4), we know that technically a coproduct Δ is a homomorphism from $\mathcal{C}$ to $\mathcal{C}\otimes\mathcal{C}$ so that only Δ_{JJ} is a coproduct. But, we will be free of language and call all Δ_{KL} as coproducts. Indeed, it is the very fact that $K\neq L$ in general in (11.38) that gives S_F^2 its "generalized" Hopf alegbra structure.

Let us now simplify the r.h.s. of (11.36). Using (11.4), (11.36) can be written as

$$\Delta\big(e^{ji}(J)\delta_J\big)_{\mu_1\mu_2,m_1m_2} = \frac{2J+1}{V}\int d\mu(g)D_{ij}^J(g)^\dagger\sum_{J'}C(K,L,J';\mu_1,\mu_2)\times$$

$$C(K,L,J';m_1,m_2)D_{\mu_1+\mu_2,m_1+m_2}^{J'},\tag{11.39}$$

with μ_1, μ_2 and m_1, m_2 being row and column indices. The r.h.s of (11.39) is

$$C(K,L,J;\mu_1,\mu_2)\, C(K,L,J;m_1,m_2)\delta_{j,\mu_1+\mu_2}\delta_{i,m_1+m_2} =$$
$$\sum_{\substack{\mu_1'+\mu_2'=j \\ m_1'+m_2'=i}} C(K,L,J;\mu_1',\mu_2')\, C(K,L,J;m_1',m_2') \times$$
$$\left(e^{\mu_1'm_1'}(K)\right)_{\mu_1 m_1} \otimes \left(e^{\mu_2'm_2'}(L)\right)_{\mu_2 m_2}. \tag{11.40}$$

Hence we have the coproduct

$$\Delta_{KL}\left(e^{ji}(J)\right) =$$
$$\sum_{\substack{\mu_1+\mu_2=j \\ m_1+m_2=i}} C(K,L,J;\mu_1,\mu_2)\, C(K,L,J;m_1,m_2)\, e^{\mu_1 m_1}(K) \otimes e^{\mu_2 m_2}(L).$$
$$\tag{11.41}$$

Writing $C(K,L,J;\mu_1,\mu_2,j) = C(K,L,J;\mu_1,\mu_2)\delta_{\mu_1+\mu_2,j}$ for the first Clebsch-Gordan coefficient, we can delete the constraint $j = \mu_1 + \mu_2$ in summation. $C(K,L,J;\mu_1,\mu_2,j)$ is an invariant tensor when μ_1,μ_2 and j are transformed appropriately by $SU(2)$. Hence (11.41) is preserved by the $SU(2)$ action on j,μ_1,μ_2. The same is the case for the $SU(2)$ action on i,m_1,m_2. In other words, the coproduct in (11.41) is equivariant with respect to both $SU(2)$ actions.

Since any $M \in Mat(2J+1)$ is $\sum_{i,j} M_{ji}e^{ji}(J)$, (11.41) gives

$$\Delta_{KL}(M) = \sum_{\substack{\mu_1,\mu_2 \\ m_1,m_2}} C(K,L,J;\mu_1,\mu_2)\, C(K,L,J;m_1,m_2)$$
$$\times M_{\mu_1+\mu_2,m_1+m_2}e^{\mu_1 m_1}(K) \otimes e^{\mu_2 m_2}(L). \tag{11.42}$$

This is the basic formula. It preserves conjugation $*$ (induced by hermitian conjugation of matrices):

$$\Delta(M^\dagger) = \Delta(M)^\dagger. \tag{11.43}$$

It is instructive to check directly that Δ_{KL} is a homomorphism, that is that $\Delta_{KL}(MN) = \Delta_{KL}(M)\Delta_{KL}(N)$. Starting from (11.41) we have

$$\Delta_{KL}\left(e^{ji}(J)\right)\Delta_{KL}\left(e^{j'i'}(J)\right) =$$
$$\sum_{\substack{\mu_1,\mu_2 \\ m_1,m_2}} \sum_{\substack{\mu_1'\mu_2' \\ m_1',m_2'}} C(K,L,J;\mu_1,\mu_2,j)\, C(K,L,J;m_1,m_2,i)$$
$$\times C(K,L,J;\mu_1',\mu_2',j')\, C(K,L,J;m_1',m_2',i')\left(e^{\mu_1 m_1}(K) \otimes e^{\mu_2 m_2}(L)\right)$$
$$\times \left(e^{\mu_1'm_1'}(K) \otimes e^{\mu_2'm_2'}(L)\right). \tag{11.44}$$

Using $(A \otimes B)(C \otimes D) = AC \otimes BD$, we have

$$\left(e^{\mu_1 m_1}(K) \otimes e^{\mu_2 m_2}(L)\right)\left(e^{\mu'_1 m'_1}(K) \otimes e^{\mu'_2 m'_2}(L)\right) =$$

$$e^{\mu_1 m_1}(K)e^{\mu'_1 m'_1}(K) \otimes e^{\mu_2 m_2}(L)e^{\mu'_2 m'_2}(L)$$

$$= \delta_{m_1 \mu'_1}\delta_{m_2 \mu'_2} e^{\mu_1 m'_1}(K) \otimes e^{\mu_2 m'_2}(L) . \qquad (11.45)$$

To get the second line in (11.45) we have made use of

$$\left(e^{\mu_1 m_1}(K)e^{\mu'_1 m'_1}(K)\right)_{\alpha\beta} = e^{\mu_1 m_1}(K)_{\alpha\gamma}e^{\mu'_1 m'_1}(K)_{\gamma\beta} = \delta_{m_1 \mu'_1}e^{\mu_1 m'_1}(K)_{\alpha\beta} .$$

$$(11.46)$$

Inserting (11.45) in (11.44) we get

$$\Delta_{KL}\left(e^{ji}(J)\right)\Delta_{KL}\left(e^{j'i'}(J)\right) =$$

$$\sum_{\substack{\mu_1,\mu_2 \\ m_1,m_2}} \sum_{\substack{\mu'_1\mu'_2 \\ m'_1,m'_2}} C(K,L,J;\mu_1,\mu_2,j)\, C(K,L,J;m_1,m_2,i)\times$$

$$C(K,L,J;\mu'_1,\mu'_2,j')\, C(K,L,J;m'_1,m'_2,i')\delta_{m_1 \mu'_1}\delta_{m_2 \mu'_2}e^{\mu_1 m'_1}(K)\otimes e^{\mu_2 m'_2}(L)$$

$$= \sum_{\substack{\mu_1,\mu_2}} \sum_{m'_1,m'_2} C(K,L,J;\mu_1,\mu_2,j)C(K,L,J;m'_1,m'_2,i')\times$$

$$\underbrace{\left(\sum_{m_1,m_2} C(K,L,J;m_1,m_2,i)C(K,L,J;m_1,m_2,j')\right)}_{=\delta_{ij'}} e^{\mu_1 m'_1}(K)\otimes e^{\mu_2 m'_2}(L)$$

$$(11.47)$$

where the orthogonality of Clebsch-Gordan coefficients is used to obtain $\delta_{ij'}$ for the factor with the under-brace. Thus,

$$\Delta_{KL}\left(e^{ji}(J)\right)\Delta_{KL}\left(e^{j'i'}(J)\right) =$$

$$\sum_{\substack{\mu_1,\mu_2 \\ m_1,m_2}} C(K,L,J;\mu_1,\mu_2,j)C(K,L,J;m'_1,m'_2,i')\delta_{ij'}e^{\mu_1 m'_1}(K) \otimes e^{\mu_2 m'_2}(L)$$

$$= \delta_{ij'}\Delta_{KL}(e^{ji'}) . \quad (11.48)$$

Upon multiplying both sides of (11.48) by the coefficients $M_{ji}N_{j'i'}$, we finally get

$$\Delta_{KL}\left(\sum_{ji} M_{ji}e^{ji}(J)\right)\Delta_{KL}\left(\sum_{j'i'} N_{j'i'}e^{j'i'}(J)\right) = \Delta_{KL}(M)\Delta_{KL}(N)$$

$$= (MN)_{ji'}\Delta_{KL}(e^{ji'}) = \Delta_{KL}(MN) , \quad (11.49)$$

as we intended to demonstrate.

It remains to record the fuzzy analogues of counit ε and antipode S. For the counit we have

$$\varepsilon\big(e^{ji}(J)\delta_J\big) = \frac{2J+1}{V} \int d\mu(g)D_{ij}^J(g)^\dagger \varepsilon(g(.)) = \frac{2J+1}{V} \int d\mu(g)D_{ij}^J(g)^\dagger 1$$

$$= \frac{2J+1}{V} \int d\mu(g)D_{ij}^J(g)^\dagger D^0(g) . \quad (11.50)$$

Using equation (11.2a) and the fact that $D^0(g)$ is a unit matrix with only one entry which we denote by 00, we have

$$\varepsilon\big(e^{ji}(J)\delta_J\big)_{00}(K) = \delta_{0J}\delta_{j0}\delta_{i0} , \quad \forall K \in \widehat{SU(2)} . \quad (11.51)$$

For the antipode, we have

$$S\big(e^{ji}(J)\delta_J\big) = \frac{2J+1}{V} \int d\mu(g)D_{ij}^J(g)^\dagger S(g(.))$$

$$= \frac{2J+1}{V} \int d\mu(g)D_{ij}^J(g)^\dagger g^{-1}(.) \quad (11.52)$$

or

$$S\big(e^{ji}(J)\delta_J\big)(K) = \frac{2J+1}{V} \int d\mu(g)D_{ij}^J(g)^\dagger D^K(g^{-1}) . \quad (11.53)$$

In an UIRR K we have $C = e^{-i\pi J_2}$ as the charge conjugation matrix. It fulfills $CD^K(g)C^{-1} = \bar{D}^K(g)$. Then since $D^K(g^{-1}) = D^K(g)^\dagger$,

$$D^K(g^{-1}) = CD^K(g)^T C^{-1} , \quad (11.54)$$

where T denotes transposition. We insert this in (11.53) and use (11.2a) to find

$$S\big(e^{ji}(J)\delta_J\big)_{k\ell}(K) = \frac{2J+1}{V} \int d\mu(g)D_{ij}^J(g)^\dagger \big(C_{ku}D^K(g)_{uv}^T C_{v\ell}^{-1}\big)$$

$$= \frac{2J+1}{V} \int d\mu(g)D_{ij}^J(g)^\dagger C_{ku}D^K(g)_{vu}C_{v\ell}^{-1}$$

$$= \delta_{JK}C_{ku}\delta_{ui}\delta_{vj}C_{v\ell}^{-1}$$

$$= \delta_{JK}C_{ki}C_{j\ell}^{-1} . \quad (11.55)$$

This can be simplified further. Since in the UIRR K,

$$\big(e^{-i\pi J_2}\big)_{ki} = \delta_{-ki}(-1)^{K+k} = \delta_{-ki}(-1)^{K-i} , \quad (11.56)$$

and $C^{-1} = C^T$, we find

$$S\big(e^{ji}(J)\delta_J\big)_{k\ell}(K) = \delta_{JK}\delta_{-ki}\delta_{-\ell j}(-1)^{2K-i-j}$$

$$= \delta_{JK}(-1)^{2J-i-j}e^{-i,-j}(J)_{k\ell} . \quad (11.57)$$

Thus

$$S\big(e^{ji}(J)\delta_J\big)(K) = \delta_{JK}(-1)^{2J-i-j}e^{-i,-j}(J) . \quad (11.58)$$

11.7 Interpretation

We recall from chapter 2 that the matrix $M \in Mat(2J + 1)$ can be interpreted as the wave function of a particle on the spatial slice $S_F^2(J)$. The Hilbert space for these wave functions is $Mat(2J + 1)$ with the scalar product given by $(M, N) = TrM^\dagger N$, $M, N \in S_F^2(J)$.

We can also regard M as a fuzzy two-dimensional Euclidean scalar field as we did earlier or even as a field on a spatial slice $S_F^2(J)$ of a three-dimensional spacetime $S_F^2(J) \times \mathbb{R}$.

Let us look at the particle interpretation. Then (11.42) gives the amplitude, up to an overall factor, for $M \in S_F^2(J)$ splitting into a superposition of wave functions on $S_F^2(K) \otimes S_F^2(L)$. It models the process where a fuzzy sphere splits into two others [134]. The overall factor is the reduced matrix element much like the reduced matrix elements in angular momentum selection rules. It is unaffected by algebraic operations on $S_F^2(J), S_F^2(K)$ or $S_F^2(L)$ and is determined by dynamics.

Now (11.42) preserves trace and scalar product:

$$Tr\Delta_{KL}(M) = TrM,$$

$$\left(\Delta_{KL}(M), \Delta_{KL}(N)\right) = (M, N).$$ (11.59)

So (11.42) is a unitary branching process. This means that the overall factor is a phase.

$\Delta_{KL}(S_F^2(J))$ has all the properties of $S_F^2(J)$. So (11.42) is also a precise rule on how $S_F^2(J)$ sits in $S_F^2(K) \otimes S_F^2(L)$. We can understand "how $\Delta_{KL}(M)$ sits" as follows. A basis for $S_F^2(K) \otimes S_F^2(L)$ is $e^{\mu_1 m_1}(K) \otimes e^{\mu_2 m_2}(L)$. We can choose another basis where left- and right-angular momenta are separately diagonal by coupling μ_1 and μ_2 to give angular momentum $\sigma \in [0, \frac{1}{2}, 1, \ldots, K + L]$, and m_1 and m_2 to give angular momentum $\tau \in [0, \frac{1}{2}, 1, \ldots, K + L]$. In this basis, $\Delta_{KL}(M)$ is zero except in the block with $\sigma = \tau = J$.

So the probability amplitude for $M \in S_F^2(J)$ splitting into $P \otimes Q \in S_F^2(K) \otimes S_F^2(L)$ for normalized wave functions is

$$phase \times Tr(P \otimes Q)^\dagger \Delta_{KL}(M).$$ (11.60)

Branching rules for different choices of M, P and Q are independent of the constant phase and can be determined.

Written in full, (11.60) is seen to be just the coupling conserving left- and right- angular momenta of $P^\dagger, Q^\dagger$ and M. That alone determines (11.60).

An observable A is a self-adjoint operator on a wave function $M \in S_F^2(J)$. Any linear operator on $S_F^2(J)$ can be written as $\sum B_\alpha^L C_\alpha^R$ where

$B_\alpha, C_\alpha \in S_F^2(J)$ and B_α^L and C_α^R act by left- and right- multiplication: $B_\alpha^L M = B_\alpha M$, $C_\alpha^R M = M C_\alpha$. Any observable on $S_F^2(J)$ has an action on its branched image $\Delta_{KL}(S_F^2(J))$:

$$\Delta_{KL}(A)\Delta_{KL}(M) := \Delta_{KL}(AM). \tag{11.61}$$

By construction, (11.61) preserves algebraic properties of operators. $\Delta_{KL}(A)$ can actually act on all of $S_F^2(K) \otimes S_F^2(L)$, but in the basis described above, it is zero on vectors with $\sigma \neq J$ and/or $\tau \neq J$.

This equation is helpful to address several physical questions. For example if M is a wave function with a definite eigenvalue for A, then $\Delta_{KL}(M)$ is a wave function with the same eigenvalue for $\Delta_{KL}(A)$. This follows from $\Delta_{KL}(BM) = \Delta_{KL}(B)\Delta_{KL}(M)$ and $\Delta_{KL}(MB) = \Delta_{KL}(M)\Delta_{KL}(B)$. Combining this with (11.59) and the other observations, we see that the mean value of $\Delta_{KL}(A)$ in $\Delta_{KL}(M)$ and of A in M are equal.

In summary, all this means that every operator on $S_F^2(J)$ is a constant of motion for the branching process (11.42).

Now suppose $R \in S_F^2(K) \otimes S_F^2(L)$ is a wave function which is not necessarily of the form $P \otimes Q$. Then we can also give a formula for the probability amplitude for finding R in the state described by M. Note that R and M live in different fuzzy spaces. The answer is

$$constant \times Tr R^\dagger \Delta_{KL}(M). \tag{11.62}$$

If M, P, Q are fields with $S_F^2(I)$ ($I = J, K, L$) a spatial slice or space-time, (11.60) is an interaction of fields on different fuzzy manifolds. It can give dynamics to the branching process of fuzzy topologies discussed above.

11.8   The Prešnajder Map

This section is somewhat disconnected from the material in the rest of the chapter.

We recall that $S_F^2(J)$ can be realized as an algebra generated by the spherical harmonics Y_{lm} ($l \leq 2J$) which are functions on the two-sphere S^2. Their product can be the coherent state $*_C$ Moyal $*_W$ product.

But we saw that $S_F^2(J)$ is isomorphic to the convolution algebra of functions D_{MN}^J on $SU(2) \simeq S^3$.

It is reasonable to wonder how functions on S^2 and S^3 get related preserving the respective algebraic properties.

The map connecting these spaces is described by a function on $SU(2) \times S^2 \approx S^3 \times S^2$ and was first introduced by Prešnajder [65, 63]. We give its definition and introduce its properties here. It generalizes to any group G.

Let a_i, $a_j^\dagger$ $(i = 1, 2)$ be Schwinger oscillators for $SU(2)$ and let us also recall that for $J = \frac{n}{2}$

$$|z, 2J\rangle = \frac{(z_i a_i^\dagger)^{2J}}{\sqrt{2J!}} |0\rangle, \qquad \sum |z_i|^2 = 1 \tag{11.63}$$

are the normalized Perelomov coherent states. If $U(g)$ is the unitary operator implementing $g \in SU(2)$ in the spin J UIRR, the Prešnajder function [65, 63] P_J is given by

$$P_J(g, \vec{n}) = \langle z, 2J | U(g) | z, 2J \rangle = D_{JJ}^J(h^{-1}gh),$$
$$\vec{n} = z^\dagger \vec{\tau} z, \quad \vec{n} \cdot \vec{n} = 1,$$
$$h = \begin{pmatrix} z_1 & -\bar{z}_2 \\ z_2 & \bar{z}_1 \end{pmatrix}. \tag{11.64}$$

Now $\vec{n} \in S^2$. As the phase change $z_i \to z_i e^{i\theta}$ does not effect P_J, besides g, it depends only on $\vec{n}$. It is a function on

$$\left(SU(2) \simeq S^3\right) \times \left[SU(2)/U(1)\right] \simeq S^3 \times S^2. \tag{11.65}$$

A basis of $SU(2)$ functions for spin J is D_{ij}^J. A basis of S^2 functions for spin J is $E^{ij}(J, .)$ where

$$E^{ij}(J, \vec{n}) = \langle z, 2J | e^{ij}(J) | z, 2J \rangle = D^J(h^{-1})_{Ji} D^J(h)_{jJ}, \quad \text{no sum on } J. \tag{11.66}$$

The transform of D_{ij}^J to $E^{ij}(J, .)$ is given by

$$E^{ij}(J, \vec{n}) = \frac{(2J + 1)}{V} \int d\mu(g) \bar{P}_J(g, \vec{n}) D_{ij}^J(g). \tag{11.67}$$

This can be inverted by constructing a function Q_J on $SU(2) \times S^2$ such that

$$\int_{S^2} d\Omega(\vec{n}) Q_J(g', \vec{n}) \bar{P}_J(g, \vec{n}) = \sum_{ij} D_{ij}^J(g') \bar{D}_{ij}^J(g), \qquad d\Omega(\vec{n}) = \frac{d\cos\theta d\varphi}{4\pi}, \tag{11.68}$$

θ and φ being the polar and azimuthal angles on S^2. Then using (2), we get

$$D_{ij}^J(g') = \int_{S^2} d\Omega(\vec{n}) Q_J(g', \vec{n}) E^{ij}(J, \vec{n}). \tag{11.69}$$

Consider first $J = \frac{1}{2}$. In that case

$$\bar{P}_{\frac{1}{2}}(g, \vec{n}) = \bar{g}_{kl} \bar{z}_k z_l = \bar{g}_{kl} \left(\frac{1 + \vec{\sigma} \cdot \vec{n}}{2}\right)_{lk} \tag{11.70}$$

where g is a 2×2 $SU(2)$ matrix and σ_i are Pauli matrices. Since

$$\int_{S^2} d\Omega(\vec{n}) n_i n_j = \frac{1}{3}\delta_{ij} \,, \tag{11.71}$$

we find

$$Q_{\frac{1}{2}}(g', \vec{n}) = Tr\tilde{g}'(1 + 3\vec{\sigma} \cdot \vec{n}) \,, \tag{11.72}$$

$$g' = 2 \times 2 \; SU(2) \; \text{matrix} \,,$$

$$\tilde{g}' = \text{transpose of} \quad g' \,.$$

For $J = \frac{n}{2}$, $D^J(g)$ acts on the symmetric product of n $\mathbb{C}^2$'s and can be written as $\underbrace{g \otimes g \otimes \cdots \otimes g}_{N\,factors}$ and (11.70) gets replaced by

$$\bar{P}_J(g, \vec{n}) = \left[Tr\bar{g}\left(\frac{1 + \vec{\sigma} \cdot \vec{n}}{2}\right)\right]^N \,. \tag{11.73}$$

Then $Q_J(g', \vec{n})$ is defined by (11.68). It exists, but we have not found a neat formula for it.

As the relation between E^{ij} and Y_{lm} can be worked out, it is possible to suitably substitute Y_{lm} for E^{ij} in these formulae.

These equations establish an isomorphism (with all the nice properties like preserving $*$ and $SU(2)$-actions) between the convolution algebra $\rho(J)$ (G^*) at spin J and the $*$-product algebra of $S_F^2(J)$. That is because we saw that $\rho(J)(G^*)$ and $S_F^2(J) \simeq Mat(2J + 1)$ are isomorphic, while it is known that $Mat(2J + 1)$ and the $*$-product algebra of S^2 at level J are isomorphic.

There are evident generalizations of P_J to other groups and their orbits.

Bibliography

[1] A. P. Balachandran, video conference course on "Fuzzy Physics", at
     http://www.phy.syr.edu/courses/Fuzzy Physics and
     http://bach.if.usp.br/FUZZY/.

[2] J. Madore, The Fuzzy Sphere, Class. Quant. Grav. **9**, 69 (1992).

[3] J. Madore, *An Introduction to Non-commutative Differential Geometry and
     its Physical Applications*, Cambridge University Press, Cambridge (1995);

[4] J. Hoppe, *Quantum Theory of a massless relativistic surface and
     a two-dimensional bound state problem*, PhD Thesis, MIT, 1982,
     http://www.aei.mpg.de/ hoppe/.

[5] A.A. Kirillov, *Encyclopedia of Mathematical Sciences*,vol 4, p.230;
     B. Kostant, *Lecture Notes in Mathematics*, vol.170, Springer-Verlag(1970),
     p.87.

[6] F. A. Berezin, General Concept of Quantization, Commun. Math. Phys. **40**,
     153 (1975).

[7] H. Grosse, C. Klimčik, P. Prešnajder, Field Theory on a Supersymmetric
     Lattice, Commun. Math. Phys., **185** (1997) 155-175 and hep-th/9507074;
     H. Grosse, C. Klimčik, P. Prešnajder, N=2 Superalgebra and Non-
     Commutative Geometry, hep-th/9603071;

[8] H. Grosse, G. Reiter, The Fuzzy Supersphere, J. Geom. and Phys., **28** (1998)
     349-383 and math-ph/9804013.

[9] A. P. Balachandran, S. Kürkçüoğlu and E. Rojas, The Star Product on the
     Fuzzy Supersphere, JHEP **0207**, 056 (2002) [arXiv:hep-th/0204170];

[10] S. Kürkçüoğlu, Ph.D. Thesis, Syracuse University, Syracuse NY, 2004.

[11] J. Ambjorn and S. Catterall, Stripes from (Noncommutative) Stars, Phys.
     Lett. B **549**, 253 (2002) [arXiv:hep-lat/0209106].

[12] See for example, P. H. Frampton *Gauge Field Theories*, The Benjamin Cum-
     mings Publishing Company, Menlo Park CA, (1987).

[13] R.D. Sorkin Int.J. Theory. Phys. **30** (1991) 923;
     A. P. Balachandran, G. Bimonte, E. Ercolessi, G. Landi, F. Lizzi, G. Sparano
     and P. Teotonio-Sobrinho, Finite Quantum Physics and Noncommutative
     Geometry Nucl. Phys. Proc. Suppl. **37C**, 20 (1995) [arXiv:hep-th/9403067].

[14] S. Minwalla, M. Van Raamsdonk and N. Seiberg, Noncommutative Pertur-

bative Dynamics, JHEP **0002**, 020 (2000) [arXiv:hep-th/9912072].

[15] B. Ydri, Non-commutative geometry as a regulator, Phys. Rev. D **63**, 025004 (2001) [arXiv:hep-th/0003232].

[16] S. Vaidya, Perturbative dynamics on fuzzy S(2) and RP(2), Phys. Lett. B **512**, 403 (2001) [arXiv:hep-th/0102212];

[17] C. S. Chu, J. Madore and H. Steinacker, Scaling limits of the fuzzy sphere at one loop, JHEP **0108**, 038 (2001) [arXiv:hep-th/0106205].

[18] B. P. Dolan, D. O'Connor and P. Prešnajder, Matrix ϕ^4 models on the fuzzy sphere and their continuum limits", JHEP **0203**, 013 (2002) [arXiv:hep-th/0109084];

[19] S. Vaidya and B. Ydri, On the origin of the UV-IR mixing in non-commutative matrix geometry, Nucl. Phys. B **671**, 401 (2003) [arXiv:hep-th/0305201].

[20] S. Vaidya and B. Ydri, New scaling limit for fuzzy spheres, arXiv:hep-th/0209131.

[21] X. Martin, A matrix phase for the phi**4 scalar field on the fuzzy sphere, JHEP **0404**, 077 (2004) [arXiv:hep-th/0402230].

[22] F. Garcia Flores, D. O'Connor and X. Martin, Simulating the scalar field on the fuzzy sphere, PoS **LAT2005**, 262 (2005) [arXiv:hep-lat/0601012].

[23] J. Medina, W. Bietenholz, F. Hofheinz and D. O'Connor, Field theory simulations on a fuzzy sphere: An alternative to the lattice, PoS **LAT2005**, 263 (2006) [arXiv:hep-lat/0509162].

[24] E. Brezin, C. Itzykson, G. Parisi and J. B. Zuber, Planar Diagrams, Commun. Math. Phys. **59**, 35 (1978).

[25] T. Azuma, S. Bal, K. Nagao and J. Nishimura, Nonperturbative studies of fuzzy spheres in a matrix model with the Chern-Simons term," JHEP **0405**, 005 (2004) [arXiv:hep-th/0401038].

[26] P. Castro-Villarreal, R. Delgadillo-Blando and B. Ydri, A gauge-invariant UV-IR mixing and the corresponding phase transition for U(1) fields on the fuzzy sphere, Nucl. Phys. B **704**, 111 (2005) [arXiv:hep-th/0405201].

[27] T. Azuma, S. Bal, K. Nagao and J. Nishimura, Perturbative versus non-perturbative dynamics of the fuzzy S**2 x S**2, JHEP **0509**, 047 (2005) [arXiv:hep-th/0506205].

[28] P. Castro-Villarreal, R. Delgadillo-Blando and B. Ydri, Quantum effective potential for U(1) fields on S(L)**2 X S(L)**2, JHEP **0509**, 066 (2005) [arXiv:hep-th/0506044].

[29] H. Grosse and A. Strohmaier, Noncommutative geometry and the regularization problem of 4D quantum field theory, Lett. Math. Phys. **48**, 163 (1999) [arXiv:hep-th/9902138].

[30] G. Alexanian, A. P. Balachandran, G. Immirzi and B. Ydri, Fuzzy CP(2), J. Geom. Phys. **42**, 28 (2002) [arXiv:hep-th/0103023].

[31] A. P. Balachandran, B. P. Dolan, J. H. Lee, X. Martin and D. O'Connor, Fuzzy complex projective spaces and their star-products, J. Geom. Phys. **43**, 184 (2002) [arXiv:hep-th/0107099].

[32] D. Karabali and V. P. Nair, Quantum Hall effect in higher dimensions, Nucl. Phys. B **641**, 533 (2002) [arXiv:hep-th/0203264];

D. Karabali and V. P. Nair, The effective action for edge states in higher dimensional quantum Hall systems, Nucl. Phys. B **679**, 427 (2004) [arXiv:hep-th/0307281];

D. Karabali and V. P. Nair, Edge states for quantum Hall droplets in higher dimensions and a generalized WZW model, Nucl. Phys. B **697**, 513 (2004) [arXiv:hep-th/0403111];

D. Karabali, V. P. Nair and S. Randjbar-Daemi, Fuzzy spaces, the M(atrix) model and the quantum Hall effect, arXiv:hep-th/0407007.

[33] J. Medina and D. O'Connor, Scalar field theory on fuzzy S^4, JHEP **0311**, 051 (2003), [arXiv:hep-th/0212170].

[34] A. Connes, *Non-commutative Geometry* San Diego, Academic Press, 1994.

[35] G. Landi, *An Introduction to Non-commutative Spaces and their Geometries* (Springer-Verlag, 1997);

[36] J.M. Gracia-Bondía, J.C. Várilly and H. Figueroa, *Elements of Non-commutative Geometry* (Birkhäuser, 2000).

[37] R. J. Szabo, Quantum field theory on non-commutative spaces, Phys. Rept. **378**, 207 (2003) [arXiv:hep-th/0109162].

[38] M. R. Douglas and N. A. Nekrasov, Non-commutative field theory, Rev. Mod. Phys. **73**, 977 (2001) [arXiv:hep-th/0106048].

[39] *Letter of Heisenberg to Peierls (1930)*, Wolfgang Pauli, Scientific Correspondence, Vol. II, p.15, Ed. Karl von Meyenn, Springer-Verlag, 1985;
Letter of Pauli to Oppenheimer (1946), Wolfgang Pauli, Scientific Correspondence, Vol. III, p.380, Ed. Karl von Meyenn, Springer-Verlag, 1993.

[40] H. J. Groenewold, On The Principles Of Elementary Quantum Mechanics, Physica **12**, 405 (1946).

[41] H. S. Snyder, Quantized Space-Time, Phys. Rev. **71**, 38 (1947); The Electromagnetic Field in Quantized Spacetime, Phys. Rev.**72** (1947) 68;

[42] C. N. Yang, On Quantized Space-Time, Phys. Rev. **72**, 874 (1947).

[43] J. E. Moyal, Quantum Mechanics As A Statistical Theory, Proc. Cambridge Phil. Soc. **45**, 99 (1949).

[44] R. Jackiw, Physical Instances of Noncommuting Coordinates, Nucl. Phys. Proc. Suppl. **108**, 30 (2002) [hep-th/0110057];

[45] A. P. Balachandran, G. Marmo, B. S. Skagerstam, A. Stern, *Classical Topology and Quantum States*, World Scientific, Singapore, 1991.

[46] D. A. Varshalovich, A. N. Moskalev and V. K. Khersonsky, *Quantum Theory of Angular Momentum*, World Scientific, New Jersey, 1998.

[47] T. Holstein and H. Primakoff, Field Dependence Of The Intrinsic Domain Magnetization Of A Ferromagnet, Phys. Rev. **58**, 1098 (1940).

[48] D. Sen, Quantum-spin-chain realizations of conformal field theories Phys. Rev. B 44, 2645(1991)

[49] H. J. Lipkin *Lie Groups for Pedestrians* , Dover Publications, 2002.

[50] P. Goddard and D. I. Olive, Kac-Moody And Virasoro Algebras In Relation To Quantum Physics, Int. J. Mod. Phys. A **1**, 303 (1986).

[51] F. Bayen, M. Flato, C. Fronsdal, A. Lichnerowicz and D. Sternheimer, Deformation Theory And Quantization. 1. Deformations Of Symplectic Structures, Annals Phys. **111**, 61 (1978);

F. Bayen, M. Flato, C. Fronsdal, A. Lichnerowicz and D. Sternheimer, Quantum Mechanics As A Deformation Of Classical Mechanics, Lett. Math. Phys. **1**, 521 (1977).

[52] H. Weyl *Gruppentheorie und Quantenmechanik The theory of groups and quantum mechanics*, New York, Dover Publications, 1950; H. Weyl, Quantum Mechanics And Group Theory, Z. Phys. **46**, 1 (1927).

[53] J. R. Klauder and B. S. Skagerstam, *Coherent States: Applications in Physics and Mathematical Physics*, World Scientific (1985).

[54] A. M. Perelomov *Generalized Coherent States and their Applications*, Springer-Verlag (1986).

[55] G. Alexanian, A. Pinzul and A. Stern, Generalized Coherent State Approach to Star Products and Applications to the Fuzzy Sphere, Nucl. Phys. B **600**, 531 (2001), [arXiv:hep-th/0010187].

[56] A. Voros, Wentzel-Kramers-Brillouin method in the Bargmann representation Phys. Rev. **A40**, 6814 (1989).

[57] R.L. Stratonovich, On distributions in representation space, Sov. Phys. JETP 4 (1957), 891-898 .

[58] J. C. Várilly and J. M. Gracia-Bondía, The Moyal representation for spin, Ann. Phys. (NY) 190 (1989), 107–148.

[59] José F. Cariñena, J. M. Gracia-Bondía and J. C. Várilly, Relativistic quantum mechanics in the Moyal representation, J. Phys. A 23 (1990), 901-933.

[60] R. Haag, *Local quantum physics : fields, particles, algebras.* Berlin, Springer-Verlag (1996).

[61] M. Kontsevich, Deformation quantization of Poisson manifolds, I, Lett. Math. Phys. **66**, 157 (2003) [arXiv:q-alg/9709040].

[62] A. B. Hammou, M. Lagraa and M. M. Sheikh-Jabbari, Coherent state induced star-product on R(lambda)**3 and the fuzzy sphere, Phys. Rev. D **66**, 025025 (2002) [arXiv:hep-th/0110291].

[63] P. Prešnajder, The origin of chiral anomaly and the non-commutative geometry, J. Math. Phys. **41**, 2789 (2000) [arXiv:hep-th/9912050];

[64] H. Grosse, C. Klimčik and P. Prešnajder, Towards finite quantum field theory in non-commutative geometry, Int. J. Theor. Phys. **35**, 231 (1996) [arXiv:hep-th/9505175].

[65] H. Grosse and P. Prešnajder, The Construction on non-commutative manifolds using coherent states, Lett. Math. Phys. **28**, 239 (1993);

[66] A. P. Balachandran and G. Immirzi, Fuzzy Nambu-Goldstone physics, Int. J. Mod. Phys. **A 18** (2003) 5981,arXiv:hep-th/0212133.

[67] M. V. Berry, Quantal Phase Factors Accompanying Adiabatic Changes, Proc. Roy. Soc. Lond. A **392**, 45 (1984).

[68] N.E. Wegge Olsen, *K-theory and C^*-Algebras-a Friendly Approach*, Oxford University Press, Oxford, 1993.

[69] U. Carow-Watamura and S. Watamura, Chirality and Dirac operator on noncommutative sphere,, Commun. Math. Phys. **183**, 365 (1997) [arXiv:hep-th/9605003].

[70] R. Jackiw and C. Rebbi, Spin From Isospin In A Gauge Theory, Phys. Rev. Lett. **36**, 1116 (1976).

[71] P. Hasenfratz and G. 't Hooft, A Fermion - Boson Puzzle In A Gauge Theory, Phys. Rev. Lett. **36**, 1119 (1976).

[72] A. P. Balachandran, A. Stern and C. G. Trahern, Nonlinear Models As Gauge Theories, Phys. Rev. D **19**, 2416 (1979).

[73] T. R. Govindarajan and E. Harikumar, O(3) sigma model with Hopf term on fuzzy sphere, Nucl. Phys. B **655**, 300 (2003) [arXiv:hep-th/0211258].

[74] Chuan-Tsung Chan, Chiang-Mei Chen, Feng-Li Lin, Hyun Seok Yang, '$\mathbb{C}P^n$ Model on Fuzzy Sphere', *Nucl.Phys.* B625 (2002) 327 and `hep-th/0105087`; Chuan-Tsung Chan, Chiang-Mei Chen, Hyun Seok Yang, "Topological Z_{N+1} Charges on Fuzzy Sphere", `hep-th/0106269`.

[75] Ludwik Dabrowski, Thomas Krajewski, Giovanni Landi, 'Some Properties of Non-linear σ-Models in Noncommutative Geometry', *Int. J. Mod. Phys.* B14 (2000) 2367 and `hep-th/0003099`.

[76] W.J. Zakrzewski, 'Low dimensional sigma models', Adam Hilger, Bristol 1997.

[77] J. A. Mignaco, C. Sigaud, A. R. da Silva and F. J. Vanhecke, 'The Connes-Lott program on the sphere', *Rev. Math. Phys.* **9** (1997) 689 and `hep-th/9611058`;
J. A. Mignaco, C. Sigaud, A. R. da Silva and F. J. Vanhecke, 'Connes-Lott model building on the two-sphere', *Rev. Math. Phys.* **13** (2001) 1 and `hep-th/9904171`.

[78] G. Landi, Projective Modules of Finite Type over the Supersphere $S^{2,2}$, Differ. Geom. Appl. **14** (2001) 95-111 and math-ph/9907020.

[79] A. M. Polyakov, Interaction Of Goldstone Particles In Two-Dimensions. Applications To Ferromagnets And Massive Yang-Mills Fields, Phys. Lett. B **59**, 79 (1975);
A. M. Polyakov and A. A. Belavin, Metastable States Of Two-Dimensional Isotropic Ferromagnets, JETP Lett. **22**, 245 (1975) [Pisma Zh. Eksp. Teor. Fiz. **22**, 503 (1975)].

[80] H. Grosse, C. Klimčik and P. Prešnajder, Topologically nontrivial field configurations in non-commutative geometry, Commun. Math. Phys. **178**, 507 (1996) [arXiv:hep-th/9510083].

[81] S. Baez, A. P. Balachandran, B. Ydri and S. Vaidya, Monopoles and solitons in fuzzy physics, Commun. Math. Phys. **208**, 787 (2000) [arXiv:hep-th/9811169].

[82] A. P. Balachandran and S. Vaidya, Instantons and chiral anomaly in fuzzy physics, Int. J. Mod. Phys. A **16**, 17 (2001) [arXiv:hep-th/9910129].

[83] H. Grosse and J. Madore, A Non-commutative version of the Schwinger model, Phys. Lett. B **283**, 218 (1992).
H. Grosse and P. Prešnajder, A non-commutative regularization of the Schwinger model, Lett. Math. Phys. **46**, 61 (1998).

[84] S. Terashima, A note on superfields and noncommutative geometry, Phys. Lett. B **482**, 276 (2000) [arXiv:hep-th/0002119].

[85] L. Bonora, M. Schnabl, M. M. Sheikh-Jabbari and A. Tomasiello, Noncommutative SO(n) and Sp(n) gauge theories, Nucl. Phys. B **589**, 461 (2000) [arXiv:hep-th/0006091];

I. Bars, M. M. Sheikh-Jabbari and M. A. Vasiliev, Noncommutative o*(N) and usp*(2N) algebras and the corresponding gauge field theories," Phys. Rev. D **64**, 086004 (2001) [arXiv:hep-th/0103209].

[86] M. Chaichian, P. Presnajder, M. M. Sheikh-Jabbari and A. Tureanu, Noncommutative standard model: Model building,Eur. Phys. J. C **29**, 413 (2003) [arXiv:hep-th/0107055];
M. Chaichian, P. Presnajder, M. M. Sheikh-Jabbari and A. Tureanu, Noncommutative gauge field theories: A no-go theorem, Phys. Lett. B **526**, 132 (2002) [arXiv:hep-th/0107037].
M. Chaichian, A. Kobakhidze and A. Tureanu, Spontaneous reduction of noncommutative gauge symmetry and model building,arXiv:hep-th/0408065.

[87] S. Kürkçüoğlu and C. Sämann, Drinfeld twist and general relativity with fuzzy spaces, [arXiv:hep-th/0606197].

[88] A. P. Balachandran, G. Marmo, N. Mukunda, J. S. Nilsson, E. C. G. Sudarshan and F. Zaccaria, Monopole Topology And The Problem Of Color, Phys. Rev. Lett. **50**, 1553 (1983);
A. P. Balachandran, G. Marmo, N. Mukunda, J. S. Nilsson, E. C. G. Sudarshan and F. Zaccaria, Nonabelian Monopoles Break Color. 1. Classical Mechanics, Phys. Rev. D **29**, 2919 (1984);
A. P. Balachandran, G. Marmo, N. Mukunda, J. S. Nilsson, E. C. G. Sudarshan and F. Zaccaria, Nonabelian Monopoles Break Color. 2. Field Theory And Quantum Mechanics, Phys. Rev. D **29**, 2936 (1984).

[89] S. Vaidya, Scalar multi-solitons on the fuzzy sphere, JHEP **0201**, 011 (2002) [arXiv:hep-th/0109102].

[90] D. Karabali, V. P. Nair and A. P. Polychronakos, Spectrum of Schroedinger field in a noncommutative magnetic monopole, Nucl. Phys. B **627**, 565 (2002) [arXiv:hep-th/0111249].

[91] N. Ishibashi, H. Kawai, Y. Kitazawa and A. Tsuchiya, A large-N reduced model as superstring, Nucl. Phys. B **498**, 467 (1997) [arXiv:hep-th/9612115].

[92] U. Carow-Watamura and S. Watamura, Noncommutative geometry and gauge theory on fuzzy sphere, Commun. Math. Phys. **212**, 395 (2000) [arXiv:hep-th/9801195].

[93] H. Steinacker, Quantized gauge theory on the fuzzy sphere as random matrix model, Nucl. Phys. B **679**, 66 (2004) [arXiv:hep-th/0307075];
H. Grosse and H. Steinacker, Finite gauge theory on fuzzy CP**2, Nucl. Phys. B **707**, 145 (2005) [arXiv:hep-th/0407089];
W. Behr, F. Meyer and H. Steinacker, Gauge theory on fuzzy S**2 x S**2 and regularization on noncommutative R**4, JHEP **0507**, 040 (2005) [arXiv:hep-th/0503041].

[94] P. H. Ginsparg and K. G. Wilson, A Remnant Of Chiral Symmetry On The Lattice, Phys. Rev. D **25**, 2649 (1982).

[95] A. P. Balachandran and G. Immirzi, The fuzzy Ginsparg-Wilson algebra: A solution of the fermion doubling problem, Phys. Rev. D **68**, 065023 (2003) [arXiv:hep-th/0301242];

[96] K. Fujikawa, Path Integral Measure For Gauge Invariant Fermion Theories,

Phys. Rev. Lett. **42**, 1195 (1979);
K. Fujikawa, Path Integral For Gauge Theories With Fermions, Phys. Rev. D **21**, 2848 (1980) [Erratum-ibid. D **22**, 1499 (1980)].

[97] A. P. Balachandran, G. Marmo, V. P. Nair and C. G. Trahern, A Nonperturbative Proof Of The Nonabelian Anomalies, Phys. Rev. D **25**, 2713 (1982).

[98] S. Randjbar-Daemi and J. A. Strathdee, On the overlap formulation of chiral gauge theory, Phys. Lett. B **348**, 543 (1995) [arXiv:hep-th/9412165];
S. Randjbar-Daemi and J. A. Strathdee, Gravitational Lorentz anomaly from the overlap formula in two-dimensions, Phys. Rev. D **51**, 6617 (1995) [arXiv:hep-th/9501012];
S. Randjbar-Daemi and J. A. Strathdee, Chiral fermions on the lattice, Nucl. Phys. B **443**, 386 (1995) [arXiv:hep-lat/9501027];
S. Randjbar-Daemi and J. A. Strathdee, Consistent and covariant anomalies in the overlap formulation of chiral gauge theories, Phys. Lett. B **402**, 134 (1997) [arXiv:hep-th/9703092];
M. Luscher, Exact chiral symmetry on the lattice and the Ginsparg-Wilson relation, Phys. Lett. B **428**, 342 (1998) [arXiv:hep-lat/9802011];
M. Luscher, Weyl fermions on the lattice and the non-abelian gauge anomaly, Nucl. Phys. B **568**, 162 (2000) [arXiv:hep-lat/9904009];
H. Neuberger, Chiral symmetry outside perturbation theory, arXiv:hep-lat/9912013;
W. Kerler, Dirac operator normality and chiral fermions, Chin. J. Phys. **38**, 623 (2000) [arXiv:hep-lat/9912022];
J. Nishimura and M. A. Vazquez-Mozo, Noncommutative chiral gauge theories on the lattice with manifest star-gauge invariance, JHEP **0108**, 033 (2001) [arXiv:hep-th/0107110].

[99] A. P. Balachandran, T. R. Govindarajan and B. Ydri, The fermion doubling problem and noncommutative geometry, Mod. Phys. Lett. A **15**, 1279 (2000) [arXiv:hep-th/9911087].
A. P. Balachandran, T. R. Govindarajan and B. Ydri, The fermion doubling problem and noncommutative geometry. II, arXiv:hep-th/000621.

[100] A. Bassetto and L. Griguolo, *Journ. of Math. Phys.* **32** (1991) 3195.

[101] H. Aoki, S. Iso and K. Nagao, Chiral anomaly on fuzzy 2-sphere, Phys. Rev. D **67**, 065018 (2003) [arXiv:hep-th/0209137].

[102] J. Madore, S. Schraml, P. Schupp and J. Wess, Gauge theory on noncommutative spaces, Eur. Phys. J. C **16**, 161 (2000) [arXiv:hep-th/0001203];
X. Calmet, B. Jurco, P. Schupp, J. Wess and M. Wohlgenannt, The standard model on non-commutative space-time, Eur. Phys. J. C **23**, 363 (2002) [arXiv:hep-ph/0111115].

[103] N. Seiberg and E. Witten, String theory and noncommutative geometry, JHEP **9909**, 032 (1999) [arXiv:hep-th/9908142].

[104] B. Ydri, Ph.D. Thesis, Syracuse University, NY,2001, arXiv:hep-th/0110006;
B. Ydri, Noncommutative chiral anomaly and the Dirac-Ginsparg-Wilson operator, JHEP **0308**, 046 (2003) [arXiv:hep-th/0211209].

[105] H. Aoki, S. Iso and K. Nagao, Ginsparg-Wilson relation, topological invariants and finite noncommutative geometry, Phys. Rev. D **67**, 085005 (2003) [arXiv:hep-th/0209223];

[106] H. Aoki, S. Iso and K. Nagao, Ginsparg-Wilson relation and 't Hooft-Polyakov monopole on fuzzy 2-sphere, Nucl. Phys. B **684**, 162 (2004) [arXiv:hep-th/0312199];

[107] H. Aoki, S. Iso, T. Maeda and K. Nagao, Dynamical generation of a nontrivial index on the fuzzy 2-sphere, Phys. Rev. D **71**, 045017 (2005) [Erratum-ibid. D **71**, 069905 (2005)] [arXiv:hep-th/0412052].

[108] M. Scheunert, W. Nahm and V. Rittenberg, Graded Lie Algebras: Generalization Of Hermitian Representations, J. Math. Phys. **18**, 146 (1977).

[109] M. Scheunert, W. Nahm and V. Rittenberg, Irreducible Representations Of The Osp(2,1) And Spl(2,1) Graded Lie Algebras, J. Math. Phys. **18**, 155 (1977).

[110] A. Pais and V. Rittenberg, Semisimple Graded Lie Algebras, J. Math. Phys. **16**, 2062 (1975) [Erratum-ibid. **17**, 598 (1976)].

[111] B. Dewitt, *Supermanifolds*, Cambridge University Press, Cambridge (1985); M. Scheunert, *The Theory of Lie Superalgebras*, Springer-Verlag, Berlin (1979).

[112] J. F. Cornwell, *Group Theory in Physics Vol. III*, Academic Press, San Diego (1989).

[113] L. Frappat, A. Sciarrino, P. Sorba, Dictionary on Lie Superalgebras, hep-th/9607161.

[114] M. Chaichian, D. Ellinas and P. Prešnajder, Path Integrals And Supercoherent States, J. Math. Phys. **32**, 3381 (1991).

[115] A. El Gradechi and L. M. Nieto, Supercoherent states, superKahler geometry and geometric quantization, Commun. Math. Phys., **175** (1996) 521, and hep-th/9403109;
A. M. El Gradechi, On the supersymplectic homogeneous superspace underlying the OSp(1/2) coherent states," J. Math Phys. **34**, 5051 (1993).

[116] M. Bordemann, M. Brischle, C. Emmrich and S. Waldmann, subalgebras with convering star products in deformation quantization: An algebraic construction for $\mathbb{C}P^n$, J. Math. Phys., **37** (1996) 6311; q-alg/9512019;
M. Bordemann, M. Brischle, C. Emmrich and S. Waldmann, Lett. Math. Phys., **36** (1996) 357;
S. Waldmann, Lett. Math. Phys., **44** (1998) 331.

[117] F. A. Berezin and V. N. Tolstoi, The Group With Grassmann Structure Uosp(1,2), Commun. Math. Phys. **78**, 409 (1981).
F. A. Berezin, *Introduction to Superanalysis*, D.Reidel Publishing Company, Dordrecht, Holland (1987).

[118] A. P. Balachandran, G.Marmo, B. S. Skagerstam and A. Stern, Supersymmetric Point Particles And Monopoles With No Strings, *Nucl. Phys.* **B164** (1980) 427;
G. Landi and G. Marmo, *Phy. Lett.* **B193** (1987) 61-66. Extensions Of Lie Superalgebras And Supersymmetric Abelian Gauge Fields,

[119] C. Fronsdal *Essays on Supersymmetry*. Mathematical Physics Studies Vol-

ume 8, Editor: Fronsdal, C. , Dordrecht, Reidel Pub.Co. 1986.

[120] J. Wess, J. Bagger *Princeton series in physics: supersymmetry and supergravity*, Princeton, Princeton University Press, 1983.

[121] A. P. Balachandran, A. Pinzul and B. Qureshi, SUSY anomalies break $N = 2$ to $N = 1$: The supersphere and the fuzzy supersphere, arXiv:hep-th/0506037.

[122] C. Klimčik, A nonperturbative regularization of the supersymmetric Schwinger model, Commun. Math. Phys. **206**, 567 (1999) [arXiv:hep-th/9903112];

C. Klimčik, An extended fuzzy supersphere and twisted chiral superfields, Commun. Math. Phys. **206**, 587 (1999) [arXiv:hep-th/9903202].

[123] S. Kürkçüoğlu, Non-linear sigma models on the fuzzy supersphere, JHEP **0403**, 062 (2004) [arXiv:hep-th/0311031].

[124] E. Witten, A Supersymmetric Form Of The Nonlinear Sigma Model In Two-Dimensions, Phys. Rev. D **16**, 2991 (1977).

[125] P. Di Vecchia and S. Ferrara, Classical Solutions In Two-Dimensional Supersymmetric Field Theories, Nucl. Phys. B **130**, 93 (1977);

[126] L. Frappat, P. Sorba and A. Sciarrino, "Dictionary on Lie superalgebras," arXiv:hep-th/9607161.

[127] C. Fronsdal, in "Essays On Supersymmetry", ed. by C. Fronsdal, D. Reidel Publishing Company, Dordrecht, 1986.

[128] M. E. Sweedler *Hopf Algebras*, W. A. Benjamin, New York, 1969. A. A. Kirillov, *Elements of the Theory of Representations*, Springer-Verlag, Berlin, 1976.

[129] G. Mack and V. Schomerus, QuasiHopf quantum symmetry in quantum theory, Nucl. Phys. B **370**, 185 (1992);

G. Mack and V. Schomerus in *New symmetry principles in Quantum Field Theory*, Edited by J.Frohlich et al., Plenum Press, New York, 1992;

G. Mack and V. Schomerus, Quantum symmetry for pedestrians, preprint, DESY-92-053.

[130] A. P. Balachandran and S. Kürkçüoğlu, Topology change for fuzzy physics: Fuzzy spaces as Hopf algebras, arXiv:hep-th/0310026.

[131] R. Figari, R. Höegh-Krohn and C. R. Nappi, Interacting relativistic boson fields in the de Sitter universe with two space-time dimensions, Commun. Math. Phys. **44**, 265 (1975).

[132] J. Pawelczyk and H. Steinacker, A quantum algebraic description of D-branes on group manifolds, Nucl. Phys. B **638**, 433 (2002) [arXiv:hep-th/0203110].

[133] A. P. Balachandran and C. G. Trahern *Lectures on Group Theory for Physicists*, Monographs and Textbooks in Physical Science, Bibliopolis, Napoli, 1984.

[134] A. P. Balachandran, E. Batista, I. P. Costa e Silva and P. Teotonio-Sobrinho, Quantum topology change in (2+1)d, Int. J. Mod. Phys. A **15**, 1629 (2000) [arXiv:hep-th/9905136];

A. P. Balachandran, E. Batista, I. P. Costa e Silva and P. Teotonio-Sobrinho, The spin-statistics connection in quantum gravity, Nucl. Phys. B **566**, 441

(2000) [arXiv:hep-th/9906174];
A. P. Balachandran, E. Batista, I. P. Costa e Silva and P. Teotonio-Sobrinho, A novel spin-statistics theorem in (2+1)d Chern-Simons gravity, Mod. Phys. Lett. A **16**, 1335 (2001) [arXiv:hep-th/0005286];

[135] S. Dascalescu, C. Nastasescu, S. Raianu *Hopf algebras : an introduction*, New York, Marcel Dekker, 2001 .

Relation of Fuzzy Physics to Brane physics have been investigated. Some articles on this subject are:

[136] A. Y. Alekseev, A. Recknagel and V. Schomerus, Non-commutative world-volume geometries: Branes on SU(2) and fuzzy spheres, JHEP **9909**, 023 (1999) [arXiv:hep-th/9908040].
A. Y. Alekseev, A. Recknagel and V. Schomerus, Open strings and non-commutative geometry of branes on group manifolds, Mod. Phys. Lett. A **16**, 325 (2001) [arXiv:hep-th/0104054].

[137] C. Klimčik and P. Severa, Open strings and D-branes in WZNW models, Nucl. Phys. B **488**, 653 (1997) [arXiv:hep-th/9609112].

[138] A. Y. Alekseev and V. Schomerus, D-branes in the WZW model, Phys. Rev. D **60**, 061901 (1999) [arXiv:hep-th/9812193].

[139] K. Gawedzki, Conformal field theory: A case study, arXiv:hep-th/9904145.

[140] H. Garcia-Compean and J. F. Plebanski, D-branes on group manifolds and deformation quantization, Nucl. Phys. B **618**, 81 (2001) [arXiv:hep-th/9907183].

[141] R. C. Myers, Dielectric-branes, JHEP **9912**, 022 (1999) [arXiv:hep-th/9910053].

[142] S. P. Trivedi and S. Vaidya, Fuzzy cosets and their gravity duals, JHEP **0009**, 041 (2000) [arXiv:hep-th/0007011].

[143] S. R. Das, S. P. Trivedi and S. Vaidya, Magnetic moments of branes and giant gravitons, JHEP **0010**, 037 (2000) [arXiv:hep-th/0008203].

Index